What This Book Is About

Traditional hand-formed, rock-burnished clay water vessel in Michoacan, Mexico.

***Water Storage* describes how to store water for home, farm, and small communities. It will help you design storage for just about any use, including fire safety and emergency, in just about any context—urban, rural, or village.**

This book includes:

- general principles to help you design, construct, and use any water system
- a look at common mistakes and how to avoid them
- how the different kinds of storage can serve you—tanks, groundwater, and ponds
- how to determine the optimum amount of storage for your needs
- how to determine the best shape and material for your storage
- how to manage aquifers sustainably for inexpensive storage of water in the ground
- plumbing details for inlets, outlets, drains, overflows, access, etc.
- storage accessories and gadgets such as automatic shutoff valves, remote level indicators, ozonators, and filters
- how to build your own high-quality tank from ferrocement
- original design innovations—published here for the first time—to improve the quality of stored water, increase water security, make maintenance easier, and reduce environmental impacts
- real-life examples of storage designs for a wide range of contexts

This book offers underlying design principles as well as design specifics. If you run into a situation not specifically covered, there's a good chance you'll be able to use these general principles to figure it out yourself.

3,500 gallon ferrocement rainwater storage urn (construction photos p. 105).

Installed water storage typically costs fifty cents to three dollars or more a gallon *($60–200/m³)*. If you've got this book in your hands, you're probably on the verge of making decisions about hundreds or thousands of dollars' worth of storage. On an average water system, this book could pay for itself a hundred times over in savings on construction and maintenance.

Most of the information otherwise available on water storage comes from vendors. Oasis Design doesn't sell water storage hardware, so you don't have to worry about being steered toward stuff you don't need. Rather, we make our living by providing information to help people have a higher quality of life with lower impact.

Wishing you best of luck with your projects,

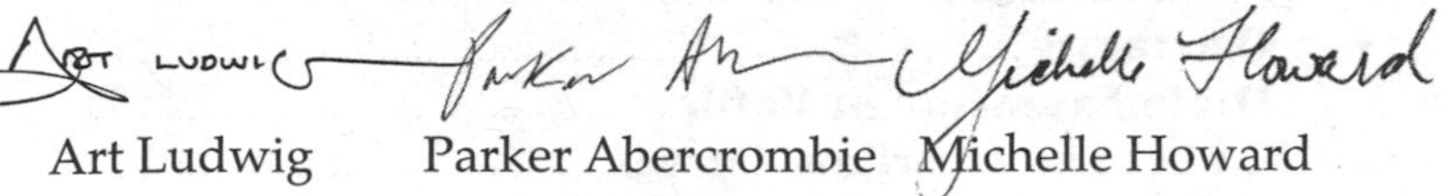

Art Ludwig   Parker Abercrombie   Michelle Howard

Contents

Water tower in California's Central Valley. Note size of person on top.

Figures and Tables

Two deep pools provide storage equal to a day's summer flow on San Jose Creek, where much of this book was written. The way this improves water quality is described on p. 11.

Chapter 1: Thinking About Water

To achieve your design goals for a water system, it is helpful to know what your goals *are*. The first order of business is to consider:

Why Store Water?

Nearly all water systems include some form of storage, most commonly a tank. Storage can be used to:

- cover peaks in demand
- smooth out variations in supply
- provide water security in case of supply interruptions or disaster
- save your home from fire
- meet legal requirements
- improve water quality
- provide thermal storage and freeze protection
- enable a smaller pipe to serve for a distant source

We're going to consider each of these reasons to store water, then look at design principles to help you frame the goals for your project.

Cover Peaks in Demand

The most common function of water storage is to cover short term use flows that are greater than the flow of the water source. For example, a tiny, one gallon-per-minute spring supplies 1440 gallons a day. This is several times more than most homes use in a day. However, almost every fixture in the home consumes water at a *faster rate* than 1 gpm while it is turned on. Even a low-flow shower head uses about 1.5 gpm.[m]

By using water stored in a tank, you can supply water to the shower faster than it is flowing from the spring. After the shower, the water will be coming in faster than it is going out, and the tank level will rise back up.

If you had a 10,000 gal tank, you could run a 100 gpm fire hose—creating the kind of blast used to bowl over hostile crowds—on the stored water from this tiny spring, for an hour and a half! Hopefully the fire would be out by then, as the tank would take several days to refill.

Short on Water

By 2025 at least 3.5 billion people—about half the world's population—will live in areas without enough water for agriculture, industry, and human needs… Worldwide, water quality conditions appear to have degraded in almost all regions with intensive agriculture and in large urban and industrial areas.

—World Resources Institute, October 2000

Smooth out Variations in Supply

In some circumstances, your storage needs will be affected by variations in the water supply. For instance, if the supply is rainwater, you will need enough storage to make it through the intervals between rainfalls. A six-month, rainless dry season requires a heck of a lot more storage than the most common kind of variable supply—a well pump that cycles on and off.

If you have a well that taps stored groundwater, a tank will save wear and tear on your pump, because the pump won't have to switch on and off every time you open a tap.

Provide Water Security in Case of Supply Interruptions or Disaster

In many places, the water supply chain from source to tap is long and made of many delicate links. If a cow steps on the supply line, a pump breaks, a wire works loose, the electricity goes out, the city misplaces your check, or there is a natural disaster, your water flow could stop. **By locating your storage as few chain links away as possible from your use point, a large measure of security is added:**

[m] *Baffled by a measurement? See Appendix A, p. 90. **Metric** units for example above: Spring supplies 3.8 lpm, or 5.5 m³/day. Low-flow showerhead uses 5.7 lpm. 38 m³ tank can supply 1400 lpm fire hose for 1.5 hrs.*

- In case of earthquake, hurricane, flood, etc., storage can be slowly emptied to meet essential needs until service is restored.
- If your well or electrical pump goes out, or your spring lines wash out, with storage you have water until you can get it fixed.

Save Your Home from Fire

Designing a system to be effective for combating fire can change its specifications radically. To put out a fire, your stored water needs to be available at a flow rate many times greater than normal. (If you want your system to have this capability, see Systems for Firefighting, p. 78.)

Meet Legal Requirements

Sometimes you may be required to install water storage simply to meet a legal requirement. On the other hand, you may be able to trade increased water storage for slack on a different legal requirement. For example, if you provide a large amount of water with good pressure that is reserved for fire emergency, a sprinkler system, and/or a hydrant, the fire department might allow you to build a narrower driveway with a smaller turnaround farther from the house than they would otherwise—thereby saving you a fortune.

Improve Water Quality

The water coming out of a properly designed tank can be of significantly higher quality than the water that goes into it. This is mostly due to attrition and settling (see Ways to Improve Water Quality in Storage, p. 10). Add an ozonator, and a tank becomes a substantial treatment step (see Ozonators, p. 71).

Provide Thermal Storage and Freeze Protection

Water has higher specific heat—stores more thermal energy per unit of weight—than any other common material. A large thermal mass of water stored within a solar greenhouse or home can help to keep it cooler in the day and warmer at night.

Also, as water changes to ice, it radiates a tremendous amount of stored energy. Imagine how much gas it would take to melt a water tank-sized ice cube. When water freezes, it releases this same amount of energy. This is why irrigating for frost protection is effective. The stored energy in water can prevent a water tank or nearby components from freezing (though in the coldest climates this may not be sufficient—see Freeze Protection, p. 73).

Evaporation consumes even more energy, which is why swamp coolers and cooling towers are effective. Water is also an effective heat transfer medium.

Finally, in rare instances it can be economical to use elevated water storage as a "battery," from which electricity is derived by running it through a hydroelectric turbine.

Enable a Smaller Pipe to Serve for a Distant Source

If the flow of your source exceeds the peak demand, you can connect to it directly without storage. However, if the source is distant, it may be cheaper to run a small pipe to nearby storage, and a big pipe from there to the use point. The small pipe would be sized to the average use, the big pipe to the peak use. The savings in materials and labor from running a smaller pipe over most of the distance can often pay for the storage and then some.

Design Principles

Oasis team hard at work in the office "annex:" Art, Michelle, and Parker. Most of the work on this book was done near this swimming hole or in our jungle of greywater- and stored rainwater-irrigated fruit trees. This oasis is surrounded at the moment by drought-dessiccated scrub, people concerned about water scarcity, and the rumble of well-drilling rigs going deeper, deeper...

This book sees water storage through the lens of integrated, ecological systems design—a view that is both global and finely detailed. You don't need to share this view to value the utility of the material in this book or benefit from its application. However, the ecological design approach offers so much advantage for so many pressing issues, I feel compelled to take a moment to look at the *design of systems* before diving into the details of water storage. (If you are not in the mood for design philosophy, please skip ahead to Separate Handling for Different Qualities of Water, p. 6.)

Ecological systems design has been my day job since 1982. My focus since 1989 has been the design of water and wastewater systems. I've designed, built, and studied water systems in 20 countries, covering many applications, in a wide range of natural and cultural conditions.

I designed my own major in ecological systems design, which I completed at UC Berkeley. However, my most important insights have occurred in the countless rugged miles I've logged exploring extraordinary wild water systems. I've become very aware of how water quality changes as it moves through natural and engineered systems. This, and my experience testing water, have led to many realizations that have influenced the designs in this book.

My approach to design—and way of life—are described in *Principles of Ecological Design*[1] a booklet I wrote in 1989. In brief, to design ecologically, follow these principles:

- **Transcend market culture**—The main obstacles to living with nature are cultural, not technical or economic. Much of American culture has been designed by marketeers, and is diametrically opposed to living cheaply and ecologically.
- **Follow nature's example**—Natural water systems are vastly more sophisticated and elegant than artificial ones. Pay them heed—they are the gold standard.
- **Intervene as little as possible**—Choose the inherently simplest solution, then implement it as well as possible. Remember that maintenance increases with number of parts, and that any moving mechanical part is itself many parts.
- **Understand that context is everything**—The context must be known in order to determine if a design is "good" or not. There are no universal solutions. There are approaches and patterns that can be applied to generate the optimum solution in a variety of contexts. Nature provides diverse solutions to all problems; she never repeats. Every branch of the tree is a different shape that fits its purpose exactly.
- **Overcome tunnel vision**—Design with a global, yet detailed view. A market economy tends to promote tunnel vision, using an exponentially increasing amount of money and resources to get higher performance on a narrow set of parameters, while the whole withers.
- **Use appropriate technology**—Cleverly matching the level of technology and power of your tools to the task at hand is cheaper, healthier, lower impact, and more enjoyable—yet ultimately more powerful than any single solution.
- **Practice moderate and efficient resource use**—Fossil fuels and electricity have severed the connection between energy source and consumer. One thin pair of wires can silently channel an unbelievable amount of energy to a pump deep underground without creating a ripple of awareness. This has enabled our relationship with energy to skew way out of scale. (Consider how you'd conserve if you were walking down 40 flights of stairs and carrying water back up in buckets.)
- **Empower users' awareness and creativity**—Design for monitoring and adjustment. We kid ourselves that our artifacts are immortal, but they all fail with time. Nature allows for failures, using the information to improve her designs and building in flexibility for changing conditions.
- **Make true progress**—Most of what is commonly called "progress" is the relocation of problems out of sight in space or time. True progress actually solves problems.

Water System Design

Here's what general ecological design principles look like when applied to the design of water systems:

- **Minimize overall negative impact on natural and social systems**—The easy way to do this is to spend as little money and use as little water as possible.
- **Create positive impacts**—For example, by slowly releasing stored water to the ecosystem during the dry season.
- **Leave as much of the water work as possible to nature.**
- **Divert water just after evaporation or soil has naturally purified it**—So that it requires no further treatment.
- **Divert water higher than the point of storage and points of use**—If possible.
- **Conserve pressure in the plumbing**—So that minimal or no pumping is required.
- **Use adequately, but not excessively, sized pipe**—To reduce friction loss where it matters, and use pipe that's as small as possible where friction loss doesn't matter.
- **Design to extract benefit from other attributes of your water**—E.g., nutrients, softness, temperature, and pressure.
- **Rigorously confine materials that are incompatible with natural cycles**—(Such as motor oil and solvents) to their own industrial cycles.
- **Add to water only materials that biodegrade into plant nutrients or non toxins**—E.g., biocompatible cleaning products.
- **Add these materials in an order that lets water cascade through multiple uses**—From those that require the cleanest water to those that tolerate the dirtiest.
- **Distribute the nutrient-laden final effluent to topsoil**—For on-site purification/reuse.

It is worth reflecting on these principles before you spend thousands of dollars on your water system.

All design is the art of trade-off. The cost-benefit curve for most design parameters shows diminishing returns. It goes up steeply at first, then levels off. However, the trade-offs and external costs continue to increase. If you push too far trying to maximize a few parameters at the expense of everything else, you will wind up with less and less total benefit. With too narrow or fuzzy a view, a myopic push to maximize the parameters in sight will inevitably compromise the parameters that are not seen (see Tunnel Vision, at right).

It is relatively easy to save water by wasting energy, or to save water and energy by wasting materials and money. It is relatively hard to make an *overall* improvement. The easiest ways to achieve this are to:

- get and stay clear on big picture goals
- change the specifications so that the full range of effects are explicitly considered
- improve the fit between subsystems
- decide what is least necessary and cut it out
- make some lifestyle accommodation

Perfection and Security Standard

A water system's perfection and security standard can be expressed as the percentage of time the system is "up." If your water is on nine days in ten, that's 90%. If your system is down three days a year, that is a perfection and security standard of 99%. If your water were off only one day every ten years, that would be a 99.97% standard.

The perfection and security standard is a relatively invisible design parameter. It is almost never discussed openly, even though it can influence the design more than just about anything. Here in America, where the culture is fear-based and where business interests dictate government policy, there tends to be runaway inflation in perfection and security standards. Our grandparents carried water in buckets from an open well in the backyard, yet we fear we'll die if the tap isn't *always* capable of delivering way more sterile water than we need, at high pressure.

Overcome Tunnel Vision with a Global, Detailed View

Tunnel vision is the number one problem behind much conventional water system development. Design tunnel vision is characterized by application of lots of resources in pursuit of narrowly defined goals, with little or no attention to the big picture. Tunnel vision is very profitable for special interests in a market economy.

Pushing legislation that mandates higher standards for a few parameters within the tunnel vision view is a diabolically effective leverage point for increasing profits at the expense of the overall good.

To optimize the overall *performance, strive for a global, yet detailed view—the essence of ecological systems design.*

FIGURE 1: TUNNEL VISION VS. GLOBAL, DETAILED VIEW

TUNNEL VISION

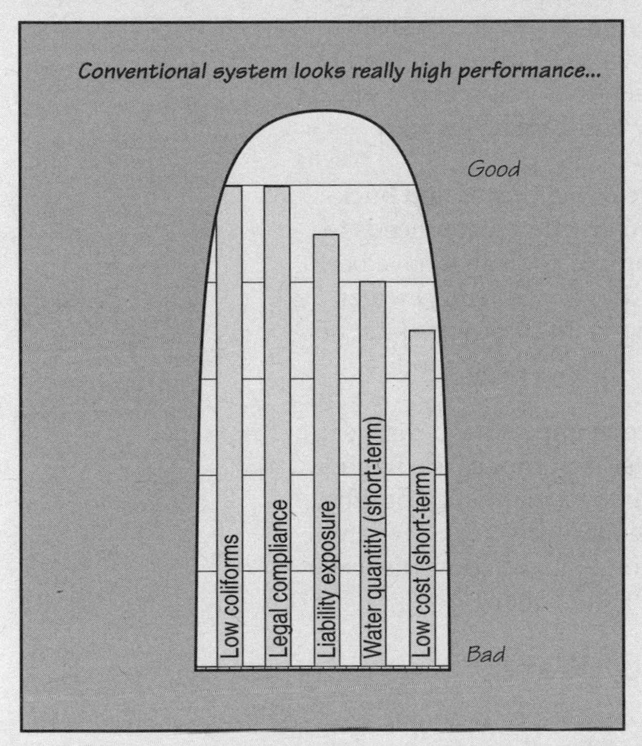

GLOBAL, DETAILED VIEW

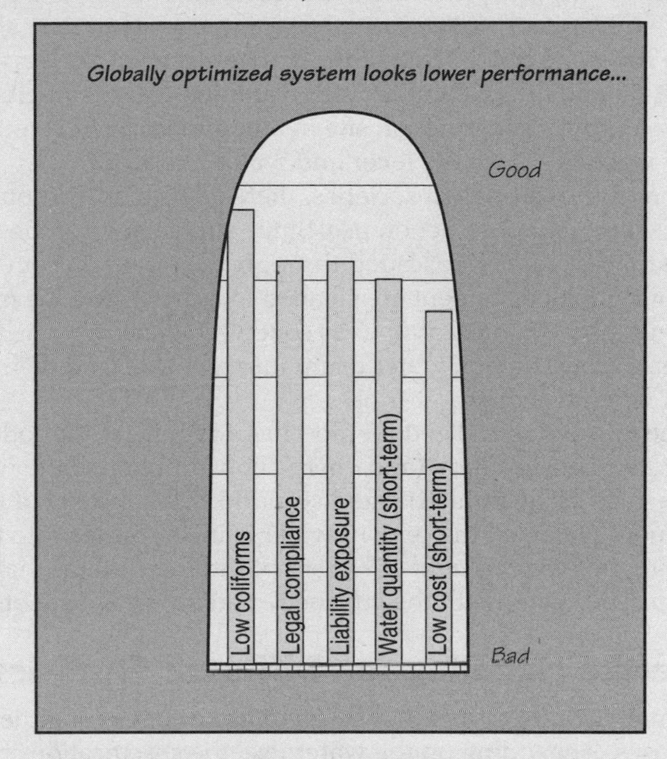

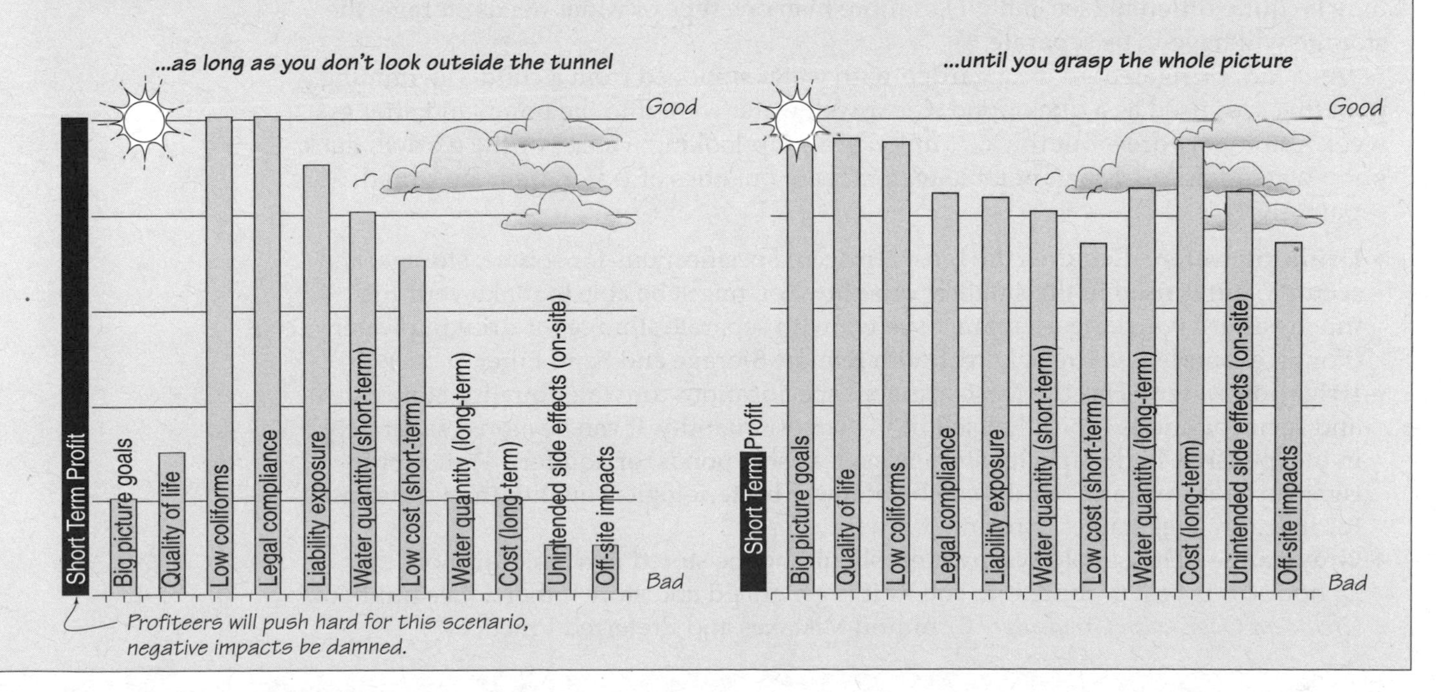

With my clients, I always haul the perfection and security standard out in the open for conscious consideration. You don't want to design for having water 100% of the time. To go from having water 95% of the time to 98%, to 99%, to 99.5% requires roughly a *doubling* of expense and environmental impact for *each* increase.

You need to take into account the consequences of insufficient storage, your emergency supply options, environmental impacts, and your budget to determine an appropriate perfection and security standard for your system. If it doesn't really matter much if you run out of water, why spend a fortune on storage? Then again, if you have a kidney dialysis machine or some other critical application, obviously you'll want to set the security standard higher.

Running Water People, Still Water People

Worldwide, people who have pressurized water on tap use about a hundred gallons a day per capita. People who carry water use about ten gallons a day to accomplish much the same tasks. Most of the difference is waste. It is easier to let a tap run than to turn it on and off. In contrast, still water just sits there serenely until you scoop it up.

On a typical construction site in an industrialized country there are hoses with spray nozzles—push the lever and water blasts out.

In non-industrialized societies, there are typically a couple of big drums and buckets of water for construction use. Since almost none of the construction water needs to be clean, the water is cascaded through various uses. For example, tools that have been used for adobe or cement are cleaned in a drum, and the muddy (or cement-y) water is then reused to make adobe (or concrete). Water for washing hands is scooped out of a "clean" construction water drum into a bucket, and then dumped into a "muddy" drum when it is dirty.

The non-industrialized method has advantages for both consumption and disposal. The water consumption (and energy consumption, if the water is pumped) is a fraction as much as in the industrialized scenario. Also, instead of leaving a giant toxic puddle of cement water, all the cement water is incorporated into the masonry work. (We are working on development of hybrid plumbing systems that combine the convenience of pressurized water with the efficiency and aesthetic benefits of still water.)

Separate Handling for Different Qualities of Water

In many contexts, it makes sense to use different qualities of water for different purposes. Depending on the water use, the specifications (including those for storage) may be quite different (see Table 1). If more than one type of water needs storage, the storage will have to be separate.

As a kid, I irrigated my first garden with water siphoned from a child's swimming pool that was used as a duck pond. Clear well water went into the pond, and after a week's storage, I drew out thick, chunky, pea soup-looking water. For the garden, duck poop water was *better* than potable water. These qualities of water often are stored separately:

- **Drinking water**—Requires the most stringent specifications for source, storage, and security, but is used in the smallest quantity. You might be able to make your life much easier by making a separate system with separate storage for drinking water. (For an example, see Creek Direct with Remote Storage and Sand Filter, p. 85.)
- **Irrigation water**—Has the least stringent specifications for water quality, storage, and security, and is typically used in the largest quantity. It can be stored separately in inexpensive, high-capacity storage such as soil, ponds, or aquifers. Water for irrigating fruit trees and shrubs can be of lower bacteriological quality than water used for irrigating vegetables consumed raw.
- **Greywater**—(Household washwater) should not be stored in a tank for more than 24 hours. It is best to route it to soil as it is generated and store it there. (See our book *Create an Oasis with Greywater* / Common Mistakes and Preferred Practices.[2])

- **Rainwater from roofs**—Is especially suited for hair washing, laundry, and flushing salts from the soil, due to its extreme softness. In the old days, inns would have a pitcher of spring water for drinking, and a separate basin of rainwater for washing.
  It is prudent to plumb rainwater downspouts to your greywater distribution or irrigation system (if you have one) so the soft rainwater can flush irrigation salts from the soil, as it soaks in and recharges the groundwater. It is not necessary to have storage to do this; it is actually most effective to do it while it is raining.
  If you have a separate rainwater harvesting tank, you can plumb it to supply the washing machine and bathtub, with any extra going to the toilet and the overflow to salt flushing. (See our forthcoming book *Rainwater Harvesting and Runoff Management.*[3])
- **Runoff water**—Generally suitable only for irrigation, flushing salts from the soil, or groundwater recharge. (See Store Water in Aquifers, p. 16, and *Rainwater Harvesting and Runoff Management.*)

TABLE 1: DIFFERENT WATER QUALITIES FOR DIFFERENT USES

	Contamination limits			
Use	**Fecal bacteria per 100 ml**	**Turbidity**	**Toxins**	**Notes**
Drinking water for sensitive humans	0	Almost none	Almost none	Needs to taste good
Well-direct groundwater recharge	0	Almost none	Almost none	
Drinking water for resistant humans	10	Low	Almost none	Needs to taste good
Drinking water for livestock	300	Moderate	Almost none	
Dishwashing water	300	Moderate	Low	
Bathing water	300	Moderate	Low	
Laundry water	1,000	Moderate	Moderate	Best if low in calcium, magnesium
Toilet flushing water	1,000	Doesn't matter	Moderate	
Irrigation of annual vegetables	1,000	Doesn't matter	Moderate	
Groundwater recharge through mulch-filled infiltration basins	3,000	Doesn't matter	Moderate	
Irrigation of fruit trees	3,000	Doesn't matter	Moderate	
Subsurface irrigation	10,000,000,000	Doesn't matter	Moderate	
Irrigation of non-fruit trees	3,000	Doesn't matter	Matters least	
Drip irrigation	3,000	Almost none	Moderate	Will clog if it has lots of solids

Warning: These figures are based on my observations of what is working in practice, and depart radically from legal standards at points—follow them at your own risk. Our other water books and articles have more information about different qualities of water for different purposes.[4]

Design Horizon

Water supply systems should typically be designed and built for a 15– to 25-year life span. The biggest variable over this time is often the population. Population determines the needed capacity for tanks, pipe sizes, etc. If long-range water demands cannot be accurately forecast, a shorter design span can be used. Note that providing abundant water tends to make the population bloom, exerting a feedback effect.

One of the best ways to account for changes in future storage needs is to provide for the addition of more storage later—for example, in additional tanks (see Figure 26, Plumbing Options for Multiple Tanks, p. 73). Instead of putting your first tank smack in the middle of the area where tanks could go, put it to one side. Later, you'll be able to add more tanks next to it if necessary.

Design for Failure, Design for Change

Another key to good water system design is to consider how the components will age, and what to do when they fail. **Every piece of the system is going to fail someday. Ask yourself what is going to happen when it does.** Will its failure be dramatic, or mundane? How long will it take to fail? Can the failed part be accessed for cleaning, repair, replacement, reuse, or recycling?

Good design makes it easy to change the system—to add another tank, a new connection, a valve, etc. As you're making the original installation, picture your future self having to come back to expand or repair the system, and make it easy to do so.

Take a look at Figure 18, Drain Options (p. 61), for example. You'll notice that the sections of PVC pipe between fittings are long enough that you could saw through them in the middle and still have enough pipe on both sides to insert a replacement valve, insert a tee to another pipe line, or make some other change or repair without having to throw anything away. If the fittings were all glued together "hub to hub," with no exposed pipe between them, to change anything you'd have to throw everything away.

Likewise, the galvanized pipe nipples in Figure 18 are the smallest size that still allows for a pipe wrench to grip the pipe. Shorter nipples are one-use; to get them out you've got to grip them with the wrench on the threads, which wrecks them.

Where the Stuff in Water Ends Up

One key to good water system design is to **focus less on the water.** Sure, the water is significant. However, there is a tendency to think that when the water is taken care of, the design is done. Most designs fail to account adequately for the other stuff in water:

- materials that sink to the bottom
- materials that float on the surface
- materials that dissolve into or out of the water
- the air that is displaced by water
- water-seeking critters that crawl or fly into the system

Take care of the other stuff, and the water will just about take care of itself. Moreover, your system will deliver higher-quality water more reliably, be less quirky, and last longer.

This is especially true if your design parameters are extreme in any way: lots of sediment and floating crud, very low pressure, barely enough water, wild fluctuations in supply or use, etc.

How do you design for the other stuff? It's easy: **Not just the water, but all the stuff that comes along with it (and the air it displaces) have to *go somewhere*.** Simply ask yourself, "Where are the air, sand, leaves, rust chunks, mineral deposits, spiders, frogs, and mosquitoes going to end up in my system?" Consider different design scenarios and you'll find the best overall disposition of the water and its companions. (See Figure 2, Common Storage Problems, p. 9.)

What Do You Have? What Can You Find?

In practice, people are likely to use a tank they already have, or the one that's sitting in the boneyard down the street, or a tank that can be purchased easily and inexpensively. This is true even if the tank is considerably smaller or bigger, of another material, or otherwise out of sync with the theoretical ideal. Cheap and easy weigh very heavily in practical terms.

Salvage storage is often economically and ecologically superior, if you can find it. One community I know purchased a used 50,000 gal tank for a good price. This tank is made much better than the brand-new, modern, galvanized tank next to it. Even though it is decades older, it looks like it is going to outlast the new tank.

How Water Quality Changes in Storage

The quality of water in natural and man-made systems is constantly changing. Every inch and every minute it is different—sometimes minutely, sometimes radically.

The water coming out of properly designed storage can be of significantly higher quality than the water that goes into it. Conversely, poorly designed storage can degrade water quality.

Changes in water quality can be physical and / or biological, intentional or unintentional. Intentional changes are accompanied by unintentional consequences. (For example, chlorination of water containing organic matter results in the formation of toxic trihalomethanes.) Some of these changes don't matter; some do.

The purpose of this section is to equip you to make sure the quality of your water improves in storage. We're going to look at:

- ways to improve water quality in storage
- hazardous disinfection byproducts
- effects of heating
- bacterial regrowth
- the problem of leaching
- water age
- how to test stored water

(There is more information on water quality under Tank Materials, p. 39, and Emergency Storage, p. 75–76.)

FIGURE 2: COMMON STORAGE PROBLEMS

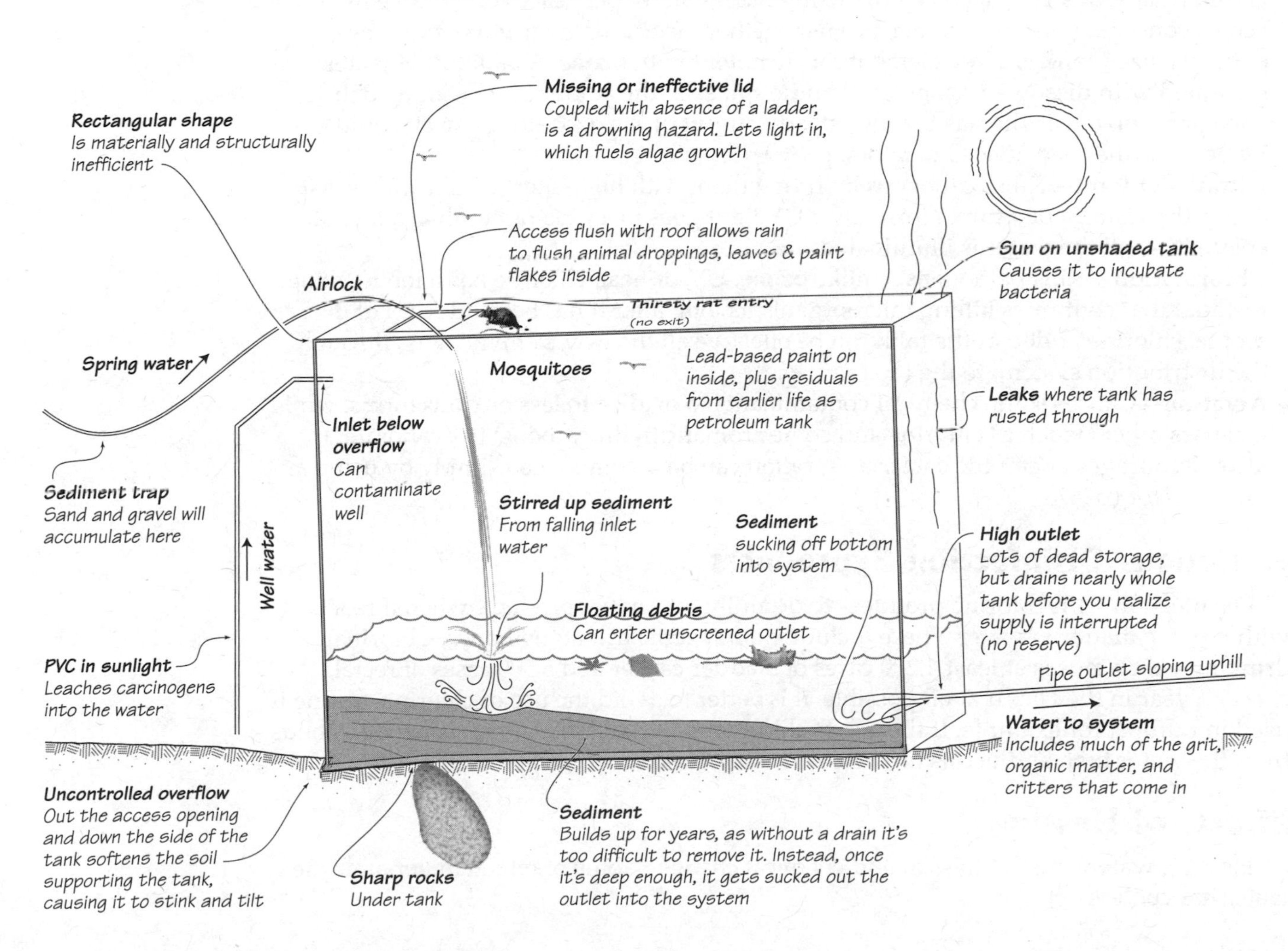

Ways to Improve Water Quality in Storage

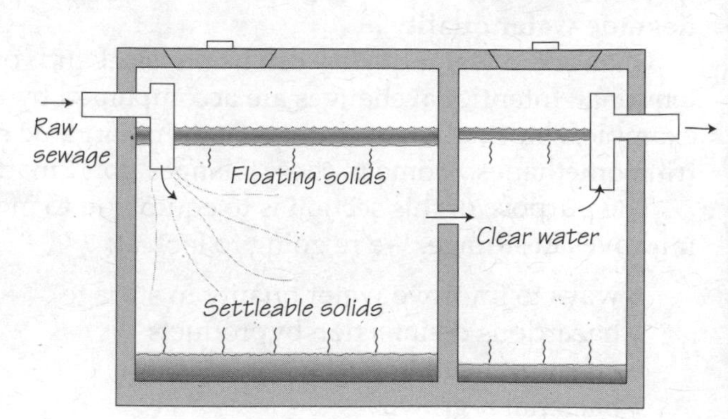

FIGURE 3: HOW SETTLING FILTERS WATER IN A SEPTIC TANK

- **Attrition**—In a well-designed container reduces harmful pathogens; they die off faster than they multiply. This is because human pathogens are designed to thrive in the human body, not a cold, nearly nutrient-free, dark water tank. The longer the water is stored and the less favorable the survival conditions, the more attrition occurs.
- **Settling**—In a still tank can reduce the amount of suspended solids (turbidity) in the water. Materials denser than water sink; materials less dense than water float. Settling for hours or days is a highly effective, low-maintenance form of filtration. This is (in part) how a septic tank turns chunky raw sewage into clarified septic effluent—water that you can often see right through. This same principle works on drinking water in a tank; it can turn it from clear to really clear. Bacteria in water can "ride" in the suspended solids, so settling can also reduce the amount of bacteria in the water column. The more still the water and the longer it sits, the more settling. The smallest, most neutrally buoyant particles will never settle. The jiggling of the water molecules themselves keeps them aloft.
- **Deflocculation**—Speeds settling. Treatment plants add a coagulant such as alum to make particles clump together and settle. This is beyond the scope of most small water systems.
- **Pasteurization**—Will kill most pathogens; for example, by heating drinking water to 149°F *(65°C)* for five minutes.
- **Ozonation**—Kills pathogens by oxidizing them with hyper-reactive, unstable molecules consisting of three oxygen atoms together instead of the usual two. An adequately sized tank is a key element for treatment with ozone. A tank full of water saturated with dissolved ozone can handle spikes in the amount of incoming debris and/or pathogens, whereas the low, steady output of the ozonator by itself could easily be overwhelmed. (See Ozonators, p. 71.)
- **Ultraviolet light**—Kills pathogens by frying them with high-energy light waves just above the visible spectrum. Normally a UV light goes in a pipe rather than a tank, to ensure that all the water is illuminated.
- **Chlorination**—Kills pathogens. Unlike ozone, UV, or heat, chlorine has a long-lasting residual that continues killing microorganisms long after it has been added. For example, chlorine added at the tank can be effective all the way from the tank, through the distribution system, to the tap.
- **Aeration**—Causes some chemical contaminants to oxidize to less noxious forms, while it causes others (such as chlorine and some aromatic hydrocarbons) to evaporate. It also discourages anaerobic bacteria. Aeration can be accomplished simply by using an inlet aerator (p. 67).

Hazardous Disinfection Byproducts

Disinfection with chlorine produces toxic, unintentional byproducts when it reacts with organic matter in water. These include carcinogenic trihalomethanes. Chlorinated drinking water causes at least 4,200 cases of bladder cancer and 6,500 cases of rectal cancer a year in the US.[5] If at all possible, it is better to avoid the use of chlorine. Ozone is used in Europe, for example. If the use of chlorine is unavoidable, filter suspended solids from the water first, and use as little chlorine as possible.

Effects of Heating

Heating water causes physical and chemical changes, some of which linger after the water has cooled:

- **Reduction in the amount of dissolved gasses**—Most significantly oxygen. Gasses are driven out of water by heating. Those little bubbles that form on the bottom of a pot long before the water boils are not steam; they are the dissolved air being driven out of solution. Water saturated with dissolved air tastes better, and is better for irrigating plants.
- **Precipitation of calcium carbonate**—On heating, calcium can precipitate out of solution, potentially clogging pipes and boilers.
- **Pasteurization**—Eliminates pathogens.

Bacterial Regrowth

Without disinfection (or sometimes with it), there can be vigorous growth of microorganisms on the inside surfaces of water infrastructure. There are numerous studies on this phenomenon,[6] but they skim over the most important question: are these bacteria a health problem?

The impression I get is that people who manage and study water systems are offended by the *idea* of bacteria in their system, whether or not they are hazardous. The most damning indictment of bacterial regrowth seems to be that their colonies could offer shelter to other, actually harmful organisms if they found their way into the system. Bacterial regrowth on its own does not appear to be a health issue.*

Water bottles that are reused over and over without washing develop vigorous bacterial growth, to the point that it can be seen and smelled. But there doesn't seem to be any indication that this is anything other than an aesthetic issue. (Studies have shown serious problems with household water containers in village settings, but the problem is not bacterial regrowth from clean water; it is unwashed hands introducing fecal matter into water containers.)

The Problem of Leaching

Natural rocks, plumbing, tanks—every material that touches your water leaches (dissolves) into it to some degree. Is it OK to drink water that has been sitting in plastic? Is it OK to drink water from a new concrete tank? These aren't questions with simple answers, especially when you reduce them to practice—you've got to contain your water in something, after all.

There is an extensive collection of information on this topic in our *Water Storage Extras*,[6] including summaries and links to dozens of studies that relate to leaching and permeation of toxins (as well as disinfection byproducts, water quality standards, and bacterial regrowth in water systems). At the risk of oversimplifying, here is a summary. Leaching is of greater concern with:

Settling and Attrition in a Swimming Hole

One hot day I went with a gaggle of young children to our local swimming hole. After hours of play, I gathered a sample from the inlet and outlet of the pool. A few days later I counted the number of colonies of general and fecal coliform bacteria in the plates.

Here's the picture: a 10′ (3 m) deep pool carved from bedrock, about 30,000 gal (110 m³) of water. A wild creek with about 30 gpm (110 lpm) of crystal clear water flows through it. The pool is looking a bit tired after a few hours of frolicking by several wild children and a couple of adults. So, which do you think was cleaner: the inlet or the outlet? (Obviously it's a trick question, or I wouldn't be asking it like this.) Well, I'll be darned if the outlet wasn't cleaner. At a loss for an explanation, I concluded that I'd switched the labels, and I re-tested. This time I tested the inlet, the surface of the water (where all the dust and stuff floats), the water column 6″ (15 cm) down, the outlet, and even the water column at the bottom, braving a cold day to dive down and uncap and recap the sample bottle.

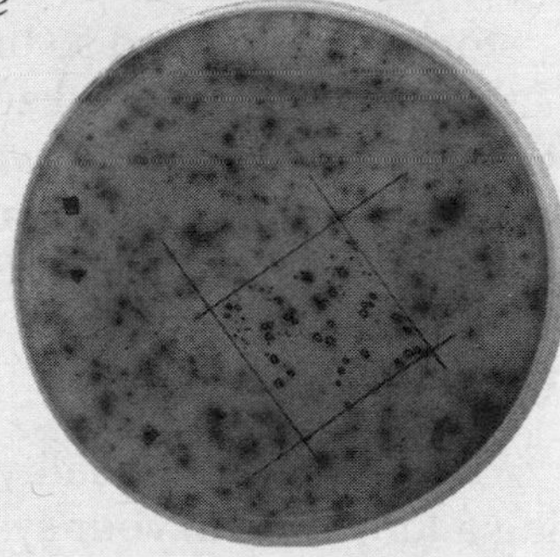
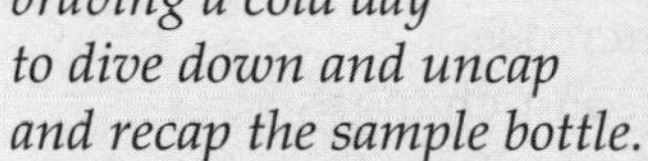

Coliscan plate

Same counter-intuitive results: the water 6″ down below the outlet was cleanest, followed by the surface, the bottom, and finally, in last place, the inlet.

I combed my memory banks for an explanation. Hmm... I remembered a set of water tests I ran at Huehuecoyotl Ecovillage in Mexico. When the rain finally came after a half-year dry season, they would usually dump the old, stale water and fill the big cistern with fresh, clean water from their waterfall. But when I tested the old and the new water, the old water was cleaner.

I think what is probably going on in our creek is that these pools, which are so big relative to the flow that if you emptied them they'd take a day to fill, had a bigger purifying effect through settling and attrition than the contaminating effect of all those (ahem) cute little butts in the water.

There is some evidence that a bacterial layer in a tank may improve water quality, the way the slime layer in a sand filter does.

- materials of greater toxicity
- materials which dissolve more readily
- longer contact time
- higher temperature
- softer water (particularly if it is less than 100 ppm total dissolved solids, like rain)
- more corrosive water (containing salt, hydrogen sulfide, etc.)
- water of extreme pH, especially low pH, acidic water

The hazard of leaching can be reduced in these ways:

- **Seek a water source that is clean to begin with**—Clean storage isn't going to help toxic water. Water providers have to publish an EPA-mandated "consumer confidence report" which describes the system's water quality. If you have your own system, consider doing your own testing. Bottled water companies are actually subject to less regulation. The Natural Resources Defense Council did a survey of bottled waters which found one particularly egregious offender was private-labeling "spring water" from a well in an industrial parking lot over a toxic waste dump![7]
- **Keep your water cool**—The colder your water, the less of your plumbing will dissolve into it. Cold also retards bacterial regrowth. Choosing a light color for your tank and/or placing it in the shade to lower its temperature is always a good idea.
- **Minimize contact time and surface area**—The leaching of plumbing materials often proceeds so slowly that reducing the contact time and surface area can significantly reduce the amount of undesirable materials leached into the water. Make sure your storage is plumbed so that water circulates through the whole water column, rather than leaving big dead zones.
- **Raise the pH of your water by adding lime.**
- **Use materials of minimum toxicity or solubility**—There is an overview of the toxicity of specific materials in Table 7, p. 40, followed by details on specific materials.

The Precautionary Principle

When an activity raises threats of harm to human health or the environment, precautionary measures should be taken even if some cause and effect relationships are not fully established scientifically.

—Wingspread conference proceedings, 1998

Scientific uncertainty + Suspected harm = Precautionary action

—NGO biotechnology briefing for the White House, 1999

Water Age

There are water quality problems that can be caused or made worse by water spending a long time in your system. These include:

- temperature increase
- taste, odor, or color changes
- decay of disinfectants
- formation of disinfectant byproducts
- bacterial regrowth/shielding of pathogens

It is preferable to design your system so that there are no stagnant backwaters where the water never turns over. In a tank, the inlet should be opposite the outlet. Abandoned runs of pipe should be capped at the beginning, not the end.

Generally, water age is not a problem in well-designed small systems if the input water is of good quality. The most common exception is rainwater harvesting or other systems that collect water with a high load of organic matter, traces of bird feces, etc. You don't want to collect this untreated water in a black tank in full sun and leave it to fester all summer, then drink it.

How to Test Stored Water

You can tell a lot about the quality of stored water by:

- **Tasting or smelling it**—Can reveal problems with over-chlorination, septic conditions, sulfur, iron, hardness, pH, and some types of leaching from plastic containers.
- **Holding it up to the light (or looking through it)**—Can reveal the amount and nature of suspended solids (turbidity), which, properly illuminated, look like dust swirling in a shaft of light.

- **Do-it-yourself lab-style tests**—Unfortunately, you can't establish that water is pathogen- or toxic chemical-free by look, smell, or taste. The conventional way to test water for pathogens is to do a few precise, expensive tests for indicator bacteria such as general and fecal coliforms, using a certified lab. However, this often does not yield an accurate picture because it is too expensive to do enough tests this way to see how the quality changes as the water moves through a system or over time.
  In our Water Quality Testing download,[4] I describe a technique for doing your own general and fecal coliform bacteria tests using materials which cost less than $2 per test (this download explains DIY tests for turbidity, flow, dissolved solids, and elevation, as well). The results aren't very precise, but you can afford to take enough samples to see how the quality changes over short distances and time spans, all throughout your system. As natural building expert Ianto Evans says: *"Better roughly right than precisely wrong."*
- **Commercial lab tests**—To test for contamination with pesticides and industrial chemicals, there isn't any alternative to sending your water to a lab.[8] If you live down current from agriculture or industry, this isn't a bad idea.

Old-fashioned drinking water filter/cooler in a farmhouse in rural Cuba. The owners pour raw water into the depression in the porous carved stone filter. From there, it drips into the clay storage urn. Louvered vents provide evaporative cooling.

Chapter 2: Ways to Store Water

We're going to take an in-depth look at water tanks in Chapters 3–5, but that doesn't mean tanks are your only—or best—storage method. Some methods will apply in a given situation while others won't. To decide on a water storage method (or methods), evaluate each option below for *your* context:

- **Source direct (no storage)**—A rarely applicable but desirable option if you have a clean source higher in elevation and flow than the water uses.
- **Store water in soil**—Inexpensive supplemental irrigation storage (not advisable in landslide areas).
- **Store water in aquifers**—Free bulk storage safe from evaporative loss, but only accessible by pump and subject to contamination and extraction by other users.
- **Store water in ponds**—Inexpensive bulk storage of water, most appropriate where rainfall exceeds evaporation and the majority of water need is for non-potable uses.
- **Store water in tanks**—Most expensive but most flexibility in location and best protection and control of the stored water.

> **Some Common Terms**
>
> *This is the way these terms are used in this book:*
>
> - ***Aquifer:*** *A geological formation saturated with water.*
> - ***Groundwater:*** *Water from an aquifer.*
> - ***Tank:*** *A large vessel for storing water, sealed except for controlled inlets, outlets, venting, etc.*
> - ***Cistern:*** *A tank without an outlet at the bottom. (Cisterns are also characterized by intermittent supply, for example, rainwater or runoff.)*
> - ***Pool:*** *A roofless tank made of any material other than a membrane or graded earth. (If the water is held by these, it's a pond.)*
> - ***Pond:*** *An artificially constructed, open-surface body of water that is supported structurally by earth, and filled at least partially by water diverted into it from elsewhere.*
> - ***Dam:*** *An impoundment astride a natural watercourse, which fills with water from the watercourse, possibly supplemented from other sources.*

We'll consider each of these options in turn. Once you've established which storage option(s) apply for you, you can skim or skip the sections which apply only to the other methods. (You can also, of course, store water in miscellaneous containers such as drums or recycled milk jugs. See Emergency Storage, p. 75, and Really Cheap Storage, p. 50.)

Source Direct (No Storage)

If the flow of the water source is equal to or greater than the peak demand, and if it is clean, reliable, and above the point of use, you don't need artificial storage. You can use the natural storage in the earth that feeds your spring or creek.

FIGURE 4: CREEK DIRECT WATER SYSTEM

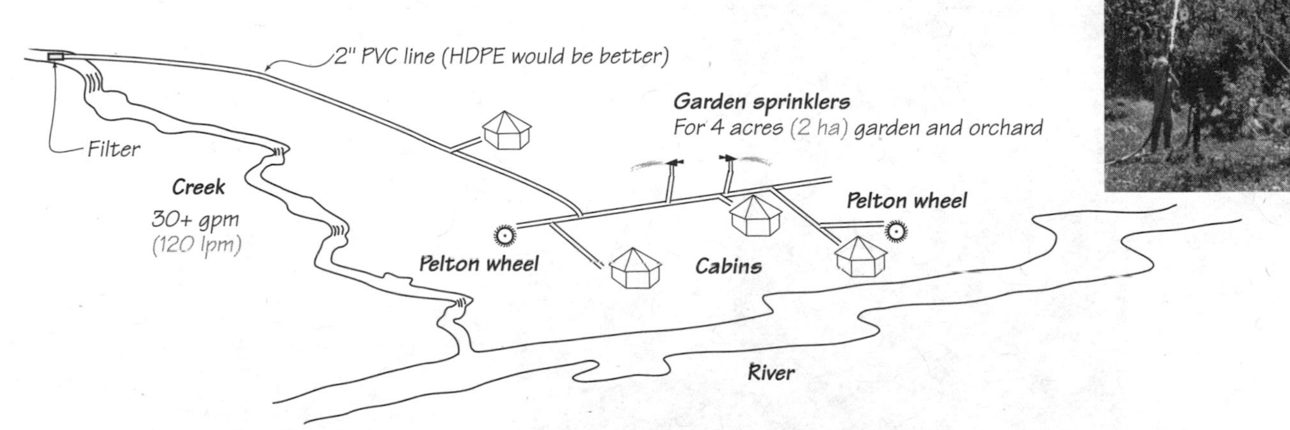

The "creek direct" system of a wilderness community in Northern California: simply a 2" pipe stuck in the creek. Can't get much simpler than that for a piped system. In the rainy season, it supplies a hydroelectric turbine or two. In summer, it still yields enough flow for all domestic use and several large sprinklers at once—and there's still water left for my son (normally confined to playing with a desert trickle at home) to go wild with the hose in the garden as long as he likes (photo above right).

The town of Stinson Beach, just north of San Francisco, has nearly a thousand people on a "creek direct" system. Thanks to abundant natural storage relative to its population, this is one of the California communities least affected by drought.

Source direct is the simplest approach to storage; there's not a lot more to say about it other than what you can see in the figures below and under Examples/Creek Direct, p. 85. The latter example features a tiny "demand-side" storage tank for days when the creek is turbid and to blast air out of the lines after servicing.

Store Water in Soil

Storing water in soil isn't going to address anything other than irrigation needs. But **for irrigation, look first to the soil to store water, and only after this has made its optimal contribution, make up the difference with other, more elaborate water storage options.**

Think like water for a moment. Visualize rain, falling on the earth. Where does it go?

- **The water from a very short, light rainfall will re-evaporate**—From the surface, with none of it cycling through plant roots.
- **A more sustained rainfall will start to fill the space between soil particles**—Plants can absorb this "field moisture" with their root hairs and pump it out through their leaves.
- **A sustained, penetrating rain will fill all the space between soil particles to "field capacity"**—At which point water will start moving downward. If it goes below root depth, it will continue on down to the groundwater (covered in the next section).
- **A variable portion of rainfall will also run off over the surface**—But that is another story. In fact, this whole process is another story, which is covered in detail in our forthcoming book *Rainwater Harvesting and Runoff Management*.[3]

Fates of Rain

- ***Transpiration***—*Productively used by plants.*
- ***Evaporation***—*A loss to the system.*
- ***Runoff***—*A loss to the system.*
- ***Infiltration***—*To soil or greywater storage.*
- ***Diversion***—*Interception by humans before one of the other fates above.*

Our concern for the moment is the water held as a film over soil particles, within root reach of the surface. While the level of an aquifer rises up and down with the amount of water in storage, soil moisture makes a thinner or thicker film of moisture over soil particles. When the space between particles isn't saturated, it's like a sponge that is not dripping. The water won't flow and can't be drawn out with a pump; it will only come out via root hairs, or slow diffusion of water as gas to the surface. Water stored in soil differs from storage in aquifers in several important ways:

TABLE 2: HOW STORAGE IN SOIL DIFFERS FROM STORAGE IN AQUIFERS

	Soil	Aquifers
Water is generally suited for	Irrigation only	All purposes
Water can be extracted by a well or spring	No or intermittently	Yes
Water can be extracted by plant roots	Yes	Only if shallow
Space between soil particles is saturated with water	Briefly	Yes
Water movement (in both soil and aquifers the flow is usually only an inch to a few feet a day)	Toward drier soil as a gas or by capillary action, downward, by bulk flow	By bulk flow, mostly lateral, seeking its level
Water has a distinct level	No	Yes, visible in well
Water has been purified of nutrients and pathogens by bacteria and roots	Not yet	Yes (unless natural purification capacity is exceeded)

The typical way to take advantage of the ability to store water in the soil in excess of current plant needs is to infiltrate rain or other excess water into a surface where plants are growing, with minimal or no runoff. There are a range of measures that can be employed to accomplish this, all to be covered in *Rainwater Harvesting and Runoff Management*.

Soil can hold on the order of an extra gallon of stored water per cubic foot,[m] between the *wilt point* and *field capacity* (the point at which the leaves of a given plant will wilt for lack of water, and the water holding capacity above which water starts to flow out of the soil like drips out the bottom of a saturated sponge). Water in the soil in excess of plants' current water needs, but less than the field capacity, will sit there until the plants need it. You can reduce "leakage" (evaporation) out of the surface of the soil by mulching, or by breaking up the surface with a hoe or plow. This breaks the capillary connection to the stored water, preventing it from evaporating.

In our tiny quarter-acre lot *(1,000 m^2)*, we can absorb 50,000 gal *(200 m^3)* of runoff diverted from the area above our land in one big storm. The water is filled with bacteria, the storage leaks out the bottom and evaporates out the top, and only a fraction of this may end up being accessed by our orchard for useful irrigation—but so what?

Storing water this way is literally dirt-cheap. The modifications to our greywater system to funnel the water to the soil cost less than $50, or about one-thousandth the cost of a tank of similar capacity. (The same end can be accomplished with zero materials cost—nothing more than furrows dug in the earth.) In any case, the captured runoff serves to flush salts below the root zone. In our orchard, water stored in soil shortens our irrigation season by two or three weeks on each end, which considerably reduces the tank size we need. What's more, the water that leaks below root reach ends up stored in an underground aquifer, from which (since our geology is favorable) we can pump it back up for use during the dry season.

Conventional management would be to shed all this excess water, then import it from somewhere else when it is needed. There are rational reasons to shed runoff this way:

- Some plants can get root rot if the soil is too waterlogged for too long.
- On a steep slope or unstable soil, you may wind up causing your house to slide down the hill or into quicksand.

Pay close attention to your context, however, and you can reap benefit from this storage with secondary benefits of reduced flooding and improved natural water quality.

In rare instances, soil storage can serve for much more. For example, the waterfall filling the pool and diversion shown on p. 25 is formed entirely from water held in the soil sponge of a small forested watershed at the top of it. This isn't an aquifer—it drains rapidly after each rain. It supplies all the water, and seasonal hydropower, for a community of a dozen or so homes. (This system, which includes dozens of tanks of every description for six months each year of zero water income, is profiled on p. 83 and on our website—search for "Huehuecoyotl.")

Store Water in Aquifers

An aquifer is an underground reservoir of groundwater. Water in an aquifer saturates all the space between particles. This saturated zone has a definite level, like an underground lake or river.

You can get water out of an aquifer with a well, or, if you're so lucky, a spring above your house, an artesian well, or a horizontal well. Water flows out by itself from these latter two as if they were springs. (See Figure 6, p. 18.)

How Water Gets into and Moves through Aquifers

Because no one ever sees an aquifer, most people have a hard time visualizing what one looks like and how water flows through it. You, however, will shortly have an above-average understanding of aquifers…

A good example to start with is the aquifer under Maruata, a Mexican village where I've worked for several years. Maruata is located in a wide river valley, with the bottom filled with sand (see photo and diagram, next page). When it rains, the rain filters down through the sand until it hits the bedrock below.

---

[m]***Metric:** an extra 130 L in each cubic meter of soil.*

Picture a kids' sandbox, with the sand slightly sloped, and one side open. Make rain by putting a sprinkler on so the sand is getting wet. Now make a highland spring, by putting a hose on trickle at the upper edge of the sand box.

Looking diagonally across the mouth of the Maruata river valley.

The water will sink into the sand. It will flow slowly under the surface, around the particles until it hits the impermeable floor, at which point it will creep toward the open end.

Underground, the water moves so much more slowly that it backs up and spreads out. It is more like a barely moving underground lake than an underground river.

If you were to dig down into the sand, you would find standing water. This is the aquifer, and you've just made a well!

If the sprinkler (rain) is coming down really hard, water may run over the surface before it all sinks in. Some of it might carry leaves and debris from the surface into the well, contaminating the groundwater. This happens in Maruata during the monsoon.

The floor and three sides of the sandbox are waterproof, like the bedrock surrounding Maruata's aquifer, while the down slope side is an open sand bank. Maruata's groundwater/underground river seeps out through an underwater sand bank into the ocean at the mouth of the valley. If I burrow my feet down into the sand while bodysurfing, I can feel the cooler, fresh, underground river water flowing out of the sand into the saltwater. If I dig a big hole in the wet sand during a falling tide and test the water at the bottom of the hole, it will be mostly fresh. If the flow of the underground river is low, and I dig during a rising tide, the water in the hole will be ocean water sloshing up into the aquifer.

If your sandbox sprinkler is on long enough, the groundwater level will rise until it reaches the surface, then run over it. In Maruata, a river runs over the surface much of the year.

Suppose you were to put a sump pump in your sandbox well and turn it on. The water level would drop rapidly, and water would start flowing toward the pump from

FIGURE 5: SIDE VIEW OF IMPACTS TO AQUIFER IN RIVER VALLEY

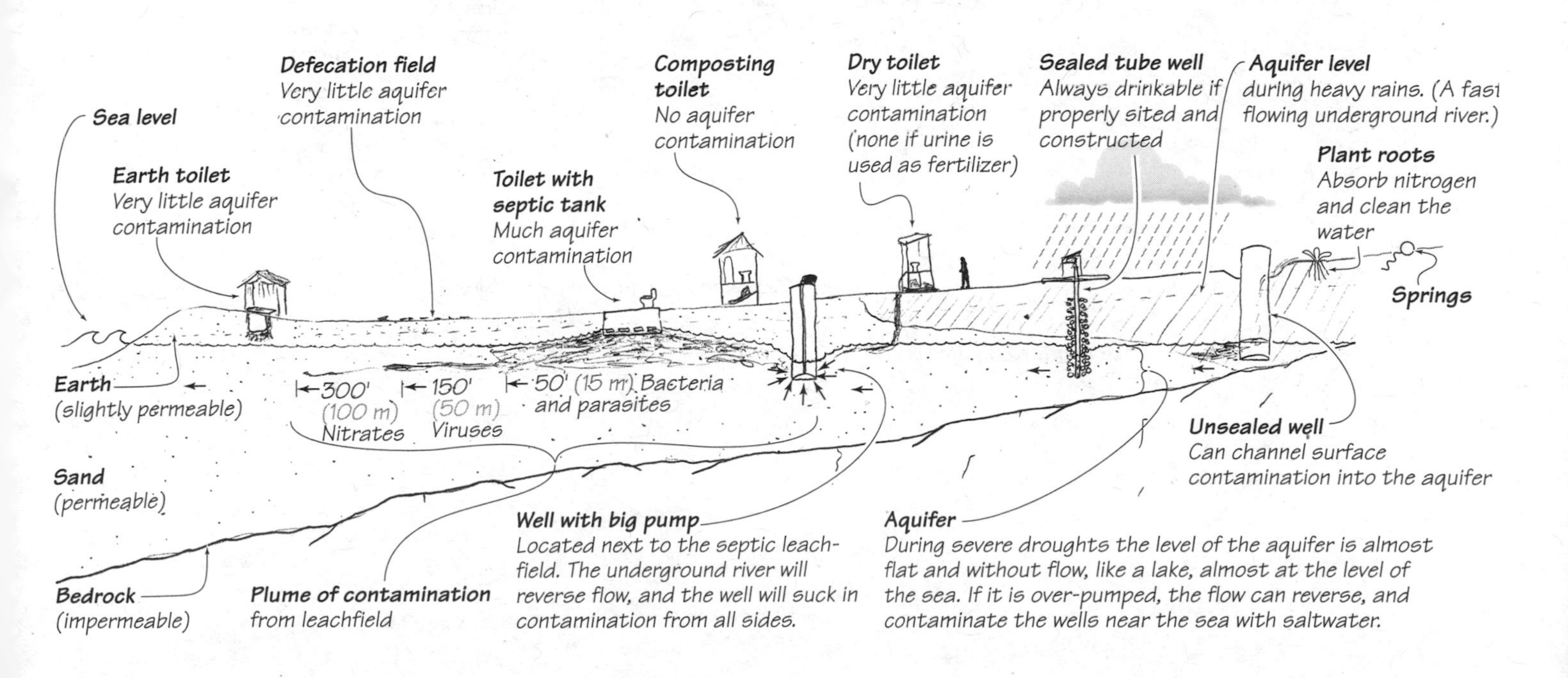

all directions, including reverse flow from the ocean toward the well. After a few minutes, the "river" on the surface would diminish, then go dry near the well, then along its whole length. If there were an ocean at the mouth of your sandbox, within a short time you'd have exhausted the freshwater and you'd be sucking saltwater into your well.

If you spray insecticide over the surface, the sprinkler will carry it down into the sandbox aquifer, and within minutes it will be coming out of your well. Ditto if you were to inject sewage below the purifying reach of plant roots. Figure 5 shows how various activities on the surface impact Maruata's aquifer. (For more on how application depth and location impacts treatment, see our other publications.[9])

Aquifers get more complicated when you add impermeable (or "confining") layers (see Figure 6). In this case, the water that seeps into the ground where you live may not end up in the groundwater directly below you. In fact, the surface that recharges your groundwater may be a long distance away, and may not correspond to the surface watershed at all. If your geology contains an impermeable layer, you may have a "perched" aquifer on top of it.

When an aquifer is sandwiched between confining layers both above and below, it can be artesian. An artesian well is pressurized. If you drill through a confining layer into a subartesian aquifer, the water will rise partway up the shaft by itself. In an artesian well, the aquifer is under enough pressure to rise up the shaft all the way and flow out the surface.

Aquifer terms

- ***Perched aquifer:*** *An aquifer "perched" atop a confining layer, with air in the soil space below it.*
- ***Artesian well (or spring):*** *A well or spring pressurized by an aquifer confined above and below by impermeable layers.*
- ***Gravity spring:*** *A spring which drains directly from saturated soil space above it, with no confining layer that could contain pressure were one to plug the spring.*
- ***Fissured aquifer:*** *Groundwater formation with bulk flow of water through cracks in bedrock.*

FIGURE 6: AQUIFER, SPRING, AND WELL TYPES

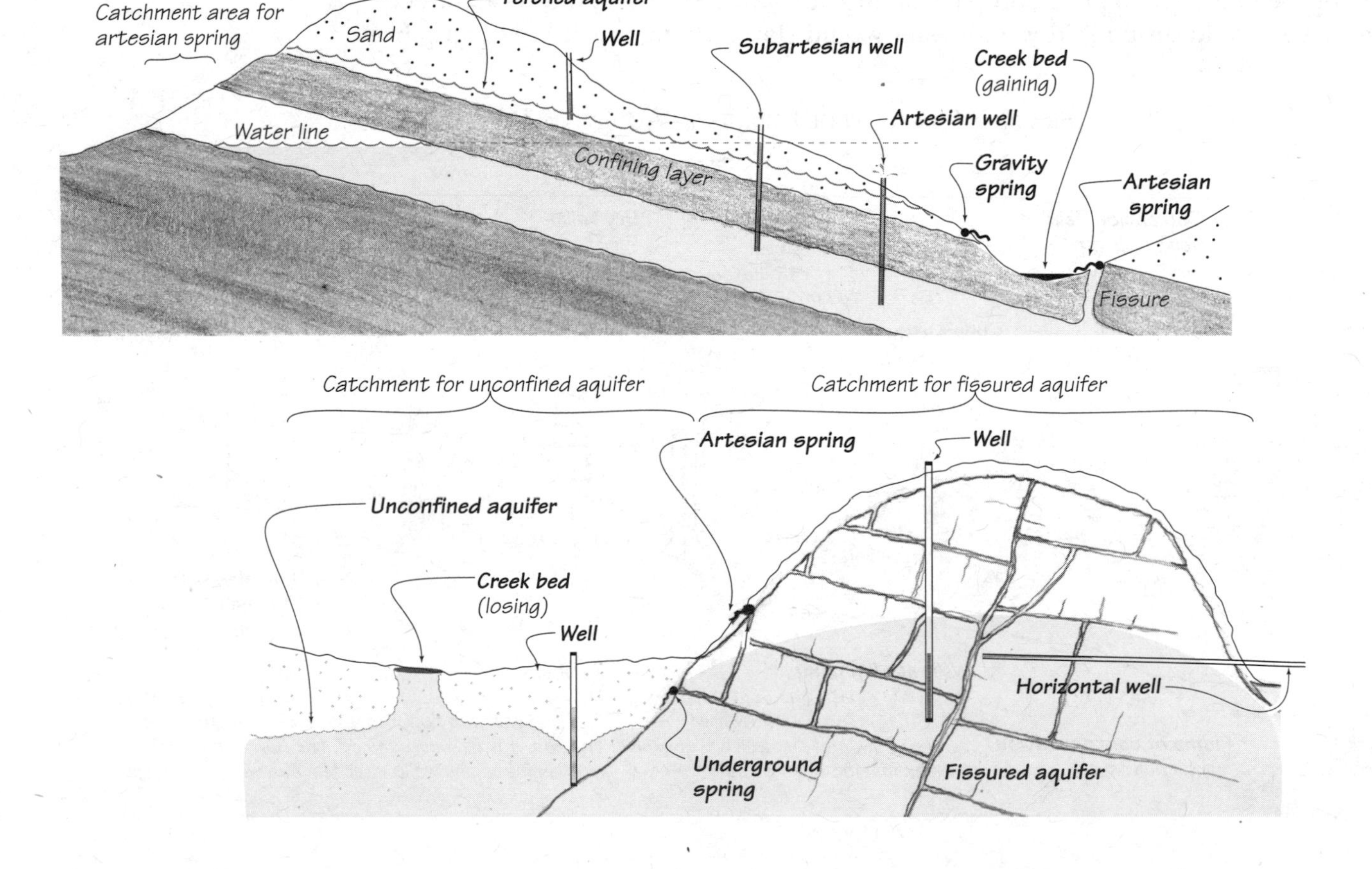

How to Increase the Amount of Water in Your Aquifer

To increase the amount of water in your aquifer:

- take less water out of it
- increase the rainfall infiltration coefficient (the percentage of rain which soaks in)
- detain water in infiltration basins
- infiltrate water through creek beds and riverbeds
- inject water into wells

Not to belabor the obvious, but conserving water is the cheapest, simplest, and lowest-impact way to relieve a water shortage, above or below ground.

If you have excess water, you can get it into storage in the ground using the strategies above. (See Runoff Harvesting Ponds, p. 21, for some info, and *Rainwater Harvesting and Runoff Management*[3] for much more on this topic.)

Aquifer Alarm

Aquifers recharge very slowly, usually at rates of 0.1 to 0.3% per year, according to the UN Environment Programme. Thus, good management and extreme restraint are required to avoid overdraft.

The Ogallala Aquifer, the largest in the world, supplies about 30% of the irrigation water in the US. Recharge rates are so low and withdrawal so high that over half the total volume of the Ogallala will be gone by the year 2020.

According to Sandra Postel, director of the Global Water Policy Project in Amherst, Massachusetts, and a senior fellow at the Worldwatch Institute, the world's farmers are racking up an annual water deficit of some 160 billion cubic meters—the amount used to produce nearly 10% of the world's grain.

"Some 40% of the world's food comes from irrigated cropland," said Postel, "and we're betting on that share to increase to feed a growing population." But the productivity of irrigation is in jeopardy from the overpumping of groundwater, the growing diversion of irrigation water to cities, and the buildup of salts in the soil.

"Our civilization is not the first to be faced with the challenge of sustaining its irrigation base. A key lesson from history is that most irrigation-based civilizations fail. As we enter the third millennium A.D., the question is: Will ours be any different?"

Conjunctive Use

"Conjunctive use" is a fancy name for adding to groundwater storage when you've got more water than you need (by surface infiltration or injection well), then taking it out when you need it. This can mean adding water during the wet season for use in the dry season, or adding water during wet decades for use during drought decades.

Conjunctive use is excellent, so long as it doesn't degenerate into the next practice…

Overdrafting, Mining Fossil Groundwater

Overdrafting groundwater is just like overdrafting your checking account—taking it out faster than it is replenished. The result is water bankruptcy. After years or decades of excess, at some point you reach the bottom. Or, you drill so deep the water is unusably hot, nasty, and salty from water-rock interactions at depth—almost as if the drill were getting too close to Hell.

Much of the world's accessible groundwater is in a state of overdraft. Overdrafting can cause permanent damage to the aquifer, subsidence of land, sinkholes, and saltwater intrusion from the ocean. With overdrafting, the best-case scenario is that you are stuck using an amount of water that corresponds to natural recharge, as you should have done in the first place. The difference is that you may be paying several times more to pump it from much farther down, possibly hundreds of feet farther down. Worse, you won't have that nice cushion of ten, a hundred, or a thousand years' past recharge in the bank for conjunctive use.

The worst-case scenario is that you were mining fossil groundwater. Fossil groundwater infiltrated into the earth in the distant past, when the climate and surface conditions were different, like when the Sahara was a dinosaur-filled swamp. With current conditions on the surface, it may not recharge even in a million years.

When mining nonrenewable fossil groundwater—as with extreme overdraft—natural recharge may be so much less than what users have gotten used to consuming that they may have to abandon the majority of their farmland, housing, or industry to adjust to the new water budget, even with state-of-the-art water efficiency.

Protecting Groundwater Quality

While tanks can be managed individually, the quality and quantity of water in aquifers depend on the wider community. Water stored in aquifers can be threatened by toxins from underground gasoline storage tanks, dry cleaners, agricultural poisons, nitrates—all the dreck of modern life which seeps down through the soil.

The best defense, of course, is to avoid such contamination in the first place. Unfortunately (or fortunately), this requires a complete rethinking of almost every aspect of our material life. While the soil is a formidable treatment engine for biological pathogens and nutrients, it is relatively transparent to artificial toxins. Toxic contamination of groundwater is increasingly widespread. Would you drink out of a parking lot gutter? If it drains into your groundwater, you are doing so already. The only reason the water isn't more gross is that modern toxins introduced decades ago are just now reaching the aquifers.

It doesn't seem that any amount of effort can truly rehabilitate an aquifer that has been contaminated by the nastier substances. Fortunately, nitrate, which is the most common groundwater contaminant, is relatively easy to flush out if the source is removed.

Saltwater intrusion is another class of threat to avoid. It is caused by overpumping, so that the groundwater level drops below that of the ocean. The remedy: pump less.

Store Water in Ponds

A pond is an artificially constructed, open-surface body of water that is supported structurally by earth, and filled by runoff, underlying springs, and/or water diverted into it from elsewhere. Ponds are generally large in capacity relative to the flow of water that maintains them.

INSTITUTE FOR SOLAR LIVING

Pond at the Institute for Solar Living, Hopland, Northern California.

Ponds or aquifers—which is better for bulk storage of excess water? Imagine setting an open-topped drum in your yard. Will it eventually fill to overflowing? If so, your climate has more rain than wind and sun. A climatologist would say your precipitation exceeds evapotranspiration. If this is the case, then ponds are an appropriate choice for water storage. If, on the other hand, the drum would never fill, then it is better to store water underground where it is protected from evaporation (see Evaporation, p. 23, for figures).

Ponds offer a lot of water storage for a low price. On the downside, they are wide open to contamination, they lose quite a lot of water to evaporation, and may leak out the bottom as well—especially new, unlined ponds. Water stored in ponds is typically of lower quality than water stored in a tank and not suited for drinking unless treated.

What's the difference between ponds and dams? A dam is an impoundment astride a year-round watercourse, whereas a pond is located where it intercepts only runoff, or is totally separated from surface water flows—its own little watershed.

Dams, particularly large ones, have serious environmental consequences. Even a small dam can create a barrier that isolates populations of fish and interrupts migrations of salmon or steelhead trout. A large reservoir of slack water behind a dam changes the temperature and sediment flow of a stream, with adverse consequences. Pooled water behind a dam is rapidly lost to evaporation. Since all dams fill to the rim with sediment within a few decades to at most several centuries, they fail any test of sustainability. We've turned the corner on dams in the US: the rate of dam removal now exceeds the rate of new dam construction. If you want to extract water or power from a stream or river, use a diversion, rather than a dam.

Properly conceived ponds, on the other hand, offer the prospect of positive environmental as well as human benefits, including:

- water storage, primarily for irrigation, livestock watering, and fire safety
- conversion of runoff to groundwater recharge or surface water that is available in dry times
- place to swim, ice-skate, fish, and view wildlife
- aquaculture, generally of fish
- aesthetic beauty
- microclimate modification (reflect low winter sun, a place to cool off in summer)
- wildlife habitat for birds, aquatic creatures, and plants
- drinking water source and hunting area for bigger creatures

Constructing a pond is a major undertaking. Start with visits to nearby ponds and questions to their owners. In many areas, government agencies such as your county's Soil and Water Conservation District, the US Department of Agriculture Natural Resource Conservation Service, or University Extension will help you design your pond for free.[10]

It is safest to retain the services of someone who knows what they are doing, with a good local track record, to help situate, design, and build the pond. There is an increasing interest in pond safety. Many states are conducting routine checks of ponds and dams, and condemning those deemed unsafe. Construct your pond well, and you'll avoid an expensive repair or liability from downstream neighbors.

The design of the pond depends on its use. Before moving earth, get clear on why you want a pond and what you'll use it for. The Ontario, Canada, Land Owner Resource Center cautions against over-ambition: "Multi-purpose ponds seldom fulfill all of their intended uses."[11]

Once you know the size of the pond and where it will be located, the amount of earthwork and the materials costs can be estimated.

Types of Man-Made Ponds and Where to Put Them

Man-made ponds have several major design variables. These yield three main classes of ponds, which we'll call Storage Ponds, Living Ponds, and Runoff Harvesting Ponds (see Table 3, next page).

- **Storage Ponds**—Essentially open, earth-supported tanks. These should be situated well out of flood plains, and, for highest water quality, should have a raised rim the whole way around them so that little or no uncontrolled runoff enters them.
- **Living Ponds**—May look just like natural ponds. They should be located out of flood plains. Sometimes they have provision for routing runoff into them or diverting it at will.
- **Runoff Harvesting Ponds**—Seasonal ponds that collect runoff that would otherwise be lost and allow it to infiltrate slowly into the soil—like a big, slow-draining infiltration basin. At their simplest, they are formed by making low dams across runoff courses. This type of pond is covered in more detail in our forthcoming book *Rainwater Harvesting and Runoff Management*.[3]

Table 3: Pond Characteristics

	Storage Ponds	Living Ponds	Runoff Harvesting Ponds
Characteristics	Open, earth-supported tanks	May look just like natural ponds	Seasonal pools of runoff—essentially big rain puddles
Lining	Usually EPDM rubber, sometimes concrete	Imported or native clay soil, or a liner covered with sand or gravel	Native soil
Plants and animals	Almost devoid of life, maybe a few mosquito fish and visiting birds	All the complexity of natural ponds, with an intricate web of plant, animal, and insect associations	Whatever can grow or live with alternating wet and dry conditions
Level fluctuation	Full to zero	Generally less than 2'	Full to zero
Depth	Deeper = less loss to evaporation	Deep enough to make cool refuge for fish, not so deep the bottom is oxygen-starved	Flexible, but generally shallow
Management and maintenance	Minimal—like a tank	High—to maintain ecological balance; weeding, stocking with fish, etc.	Minimal—monitor during big rain events
Water quality	Nearly drinkable	Low, often turbid, and full of free-floating algae	Low, often turbid, and full of free-floating algae, tannins
Water source	Filled from an external source. Runoff is generally excluded	Filled from an external source and/or underlying springs. Runoff captured, excluded, or divertable	Filled entirely with captured runoff
Cost	High	High	Low
Uses			
Water storage	Whole volume	Top 2' only	Whole volume—if any
Groundwater recharge	No	Maybe	Yes
Fishing	Maybe	Yes	Maybe, if full long enough
Swimming	Yes, but aesthetics are often lacking	Yes, but pier or dive platform may be desirable to avoid mucky bottom	Seasonally, but often shallow and mucky
Wildlife benefit	Minimal	Considerable	Considerable
Aquaculture	Maybe	Yes	If full long enough. Can plant land crops as water recedes
Typical accessories	Chain link fence	Pier, diving platform	Laundry washboard

Construction techniques go hand-in-hand with pond location:

- **Excavation ponds**—Made by digging in flat land and piling the earth into levees around the hole. In areas with high groundwater, the hole may fill with water by itself, in which case an impermeable bottom is not required.
- **Embankment ponds**—Formed by making a levee between two hillsides. Don't build this kind of pond astride a permanent watercourse or large-volume runoff channel, as the likelihood of ecological damage and failure is too high.
- **Combination ponds**—Made by a combination of the above two techniques, i.e., cut and fill.

Figure 7: Pond Construction Classes—Excavated, Embankment, and Combination

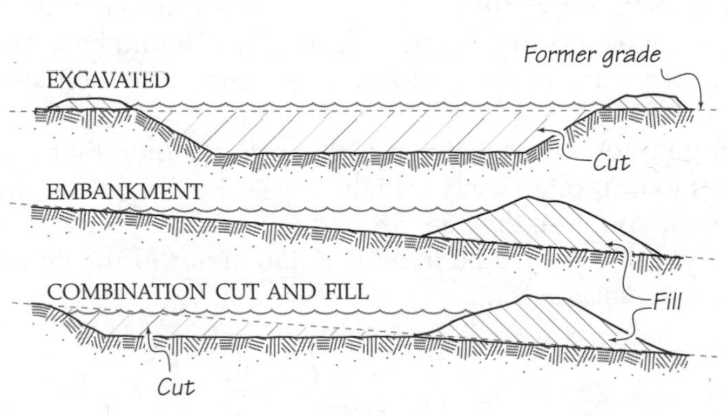

Flat sites are generally considered easiest for pond construction. However, natural folds in the land can "hold" a pond in a pleasing way, and possibly reduce the amount of costly re-arranging of soil. The pond needs to be built where there is access for heavy machinery.

Native soil that holds water is clearly a positive site attribute for a pond. Fissured bedrock may drain a pond. Maintaining a pond in a swampy area may be expensive and problematic.

Clearly, it is advantageous to situate your pond such that it can fill by gravity from the water source. Vegetation around the pond site will reduce erosion and improve water quality. Convenient access and privacy are other considerations.

A concrete-lined storage pond excavated into the spine of a ridge above an avocado orchard. It is filled by water pumped up from a creek diversion through a few hundred feet of 4" pipe.

Pond Water Sources

The source requirements vary by the pond type:

- **Storage ponds**—Require a water source of quality commensurate with the end use, just like a tank, but with a bit of extra quantity to compensate for evaporation.
- **Living ponds**—Require water of a quality that will support aquatic life, in a quantity that can maintain the pond level constant year round, despite evaporation, leakage out the bottom, and any extraction of water for other uses. Living ponds are often maintained by a diversion from a natural surface watercourse, with provision for blocking the entry of storm water and sediment.
- **Runoff harvesting ponds**—At their simplest are formed by making low levees across runoff courses. Quality generally isn't an issue. The quantity shouldn't blow out the dam.

INSTITUTE FOR SOLAR LIVING

A living pond.

You don't want too much water. Flooding causes problems in sport fishing ponds, in particular. It is good to be able to control the entry of water, sediment, and nuisance fish into the pond with a screen and/or valves.

For aquaculture, the water should be tested. Runoff can contain agricultural chemicals; groundwater can contain nitrate or carbon dioxide. Groundwater from springs or runoff from naturally forested areas are superior water sources.

In areas where ponds are common, you can probably find a locally known formula for how much area of watershed it takes to keep a pond of a given volume full. For example, in Virginia, it takes about 3 acres of watershed to maintain an acre-foot of pond volume, more if the watershed is forested with sandy soil, less if it is pasture or clay soil.[m, 12]

Runoff-harvesting storage pond for a village in Mexico. This EPDM-lined pond is filled through a mile (1.5 km) of 2.5" HDPE line from a seasonal creek. It fills for about four months and is used primarily for laundry during the long dry season.

Evaporation

How much water will evaporate? That depends on the surface area, heat, humidity, and wind. These values can often be found for a nearby weather station, as well as direct evaporation figures. Evaporation values for a dry month vary from a few inches in temperate climates to half a dozen in full desert conditions.[13] For even a modest pond, the amount of evaporative water loss is considerable. For example, a quarter-acre pond evaporates about 4" of water in an average dry month. That equals a water loss of 26,000 gallons.

[m] ***Metric:*** *In Virginia, it takes 12,000 m² to maintain 1,200 m³ of pond volume.*

Average evaporation from the 3,000 acre surface of Lake Cachuma, a reservoir above Santa Barbara, California, exceeds *10,000 gpm,* for every minute of the year—a powerful reminder that in arid zones it is preferable to store water underground.[m] (See Water Storage Extras[6] to calculate evaporation loss for your pond.)

Pond Size

To save on construction costs, maintenance effort, and environmental impacts, the pond should be no bigger than necessary for your intended use. Smaller ponds have fewer problems with wind wave erosion on the banks. Bigger ponds store more water at lower unit cost.

- **Storage ponds**—Can be sized like tanks (see p. 32).
- **Living ponds**—Should be no bigger than the water source can comfortably support. For sport fishing, ponds should be 1 acre or bigger *(4,000 m²)*, to provide sufficient cover and food for a population of fish that can't easily be depleted.
- **Runoff harvesting ponds**—Usually sized to the topography and for ease of construction. The upper size limit is the size at which they can contain all the runoff they are likely to encounter.

Pond Depth

Ideal pond depth is a function of the intended use and the weather conditions:

- **For water storage**—Deeper is better. Ponds should be at least 12 feet *(3.5 m)* deep, up to a maximum of about 20' *(6 m)*.
- **For living ponds**—Depth should be between 8' and 15' *(2.5–4.5 m)* in 25% of their basin. The colder the climate, the deeper the pond, up to the maximum of about 15' *(4.5 m)*. Beyond this depth, there may be a deep zone without oxygen, which can mix with the upper layers and kill fish if the pond "turns over," i.e., the water layers swap position due to a change in relative temperature.
- **Runoff harvesting ponds**—Typically shallow: a few to 5' or 6' deep *(0.75–2 m)*. If they are in watercourses, the risk of catastrophic failure with higher walls is too great.

Pond Shape

While round is the most efficient pond shape, a natural shape is most pleasing. A kidney shape is a popular compromise. Deep water along the shore discourages weeds. The inside walls of the pond should slope about 1:2, the outside of the walls 1:2 or 1:3.

Pond Inlets and Outlets

The pond should have:

- **A permanent drain**—Capable of draining the pond in five days or less.
- **A controllable inlet**—To exclude floodwaters, sediment, and unwanted fish.
- **A fence to exclude livestock**—They'll make a mess of the water. Instead, provide a watering trough supplied by the pond outside the fence.
- **A well-armored overflow**—Generously sized so that overflow goes through it rather than over the unprotected face of the levee. Overflows are often cut into undisturbed native soil beyond the sides of the levee. If the overflow goes over the fill soil of the levee, it should be well-armored against erosion. Ponds placed in natural runoff courses must be designed to withstand floodwaters.

Pond Liners

Ponds may be lined with native clay soil, sandy soil sealed with bentonite clay, or a welded liner of EPDM at least 0.08" *(2 mm)* thick. Rubber liners should be covered with at least 6" *(15 cm)* of sand or fine soil to protect them from punctures (although water storage ponds which are not used for any other purpose often have just the naked liner).

---

[m]**Metric:** *Evaporation ranges from 5–20 cm/month. A 1,000 m³ pond evaporates about 10 cm or 100 m³ in an average month. 1,200 ha Lake Cachuma evaporates 38 m³ a minute!*

Overflow of Runoff Diversion and Storage Pool Huehuecoyotl Ecovillage

***Tongue-shaped overflow** on an open, 12,000 gal (44 m^3) runoff diversion dam. The smooth curves make it more self-cleaning of leaves when flow is a trickle, keeping most floating debris out of the system. The wide opening increases the capacity, enabling the overflow to be placed higher and getting a few extra inches of capacity out of the dam. It replaced a 4" pipe, which clogged repeatedly. At peak flow, the water runs over several feet (2 m) of the top of the dam wall.*

***The diversion dam** is made of mortared stone, the preferred material for a diversion in a watercourse. This pool has zero wildlife cost: this watercourse is bone dry six months a year. The pool extends the availability of surface water for wildlife a month or two.*

Levee Construction

The levee (wall) contains the pond water. When building a levee, as with pond construction in general, you should retain the services of an expert.

We caution on environmental grounds against damming a year-round watercourse. These type of ponds often fail structurally—another reason to make "excavation" ponds instead, where soil is removed and piled up to make levees outside of a watercourse. It is not advisable to make a levee higher than 20' *(6 m)*.

Once you know the size of the pond and where it is going, the amount of earthwork and the materials costs can be estimated.

The site must be cleared of vegetation and topsoil, and the excavation marked with stakes.

The joint between the levee and the native soil is a weak spot, which should be armored against leakage with a clay core (see Figure 8, next page).

After filling the core trench with clay, the drain line (with anti-seep collars) can be installed.

The levee is constructed in thin lifts of well-compacted clay soil, and needs to be wide enough at the top that creatures such as muskrats can't bore extra outlets through it.

The spillway is where floodwaters that exceed the capacity of the overflow can escape. It is the most likely point of failure on a pond. It should be sized generously, and if possible be situated in undisturbed soil to the side of the levee, instead of pouring down the face of it. In the latter case, armor it against erosion.

The Virginia Extension[12] uses this formula to size spillways: Add 15' to half of the watershed area in acres. For example, a 50 acre watershed should have 40' of spillway, 200 acres 115'. This should result in overflows less than a foot high, which will reduce the loss of sport fish and reduce the forces on the spillway.[m]

The sides of the levee can be armored with rock riprap against wave action.

[m]***Metric:** Multiply watershed area in hectares by 3.7 and add 4.6 m to get spillway width.*

Figure 8: Levee Cross Section

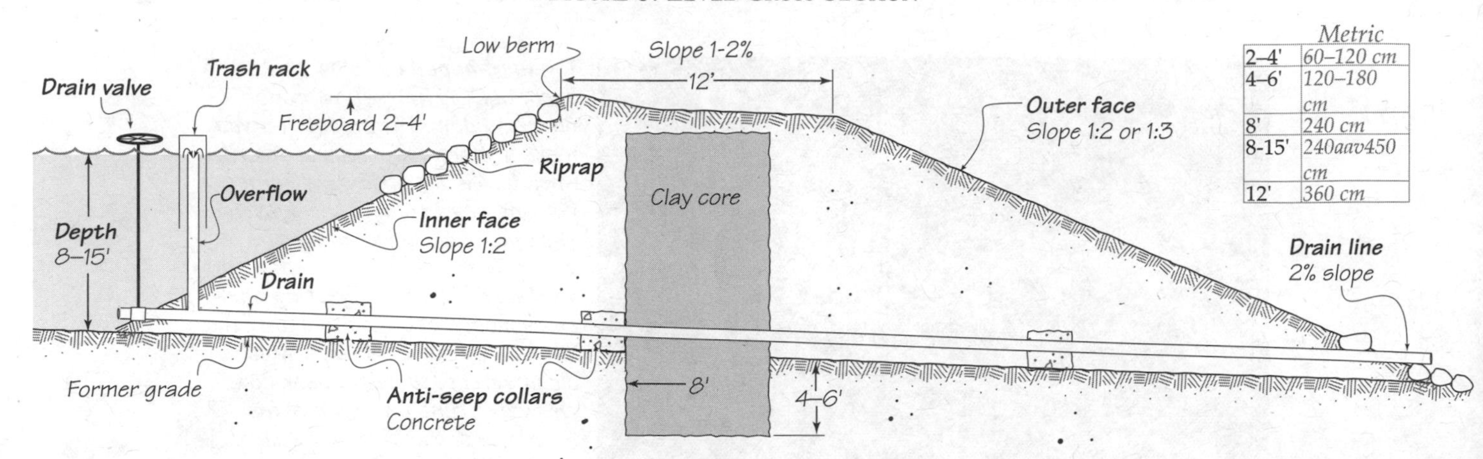

Construction of a 100' (30 m) diameter embankment pond in the Cuyama Valley, California. It is filled by gravity with water from a creek diversion, through a mile of 2" pipe. It is used for swimming and emergency storage.

Wildlife and Ponds

A pond will always attract wildlife. If you want to attract more critters, or specific species, you can attract them with habitat and a variety of fruit-bearing plants. A shallower, irregular shoreline with varied cover and open exposures will attract more creatures, as will nesting boxes.

Sport Fish in Ponds

Water characteristics must be right for a particular species of fish to thrive:[13]

- **Water temperature**—Is critical. Cold water fish, such as trout, require surface water temperatures of 60–70°F *(15–20°C)*. Cool water fish such as smallmouth bass, pike, and rock bass can tolerate temperatures of 70–80°F *(20–26°C)*. If you expect higher temperatures, stock with warm water fish such as catfish, bluegill, and largemouth bass, which can tolerate temperatures of 90°F *(32°C)*.
- **The pH**—Should be between 5 and 10. To raise pH, powdered lime can be added.
- **Dissolved oxygen**—Crucial for fish. Most of the oxygen in ponds comes from photosynthesis by water plants. There should be enough water plants to generate oxygen during the day, but not so much rotting, matted algae and root tangles that the oxygen is consumed overnight. Oxygen levels of 5–6 ppm are required for fish to thrive.
- **Water clarity**—Helps sport fish. A clear pond will support significantly more of them than a muddy pond.

Livestock cause all kinds of water quality problems and should be fenced out of sport fishing ponds.

Pond Maintenance

Levees must be protected from the tunneling and digging of crayfish, muskrats, beavers, and nuisance fish. Aquatic weeds may need to be controlled—this can often be accomplished by lowering the water level and removing the weakened or killed plants. The pond may periodically need to be dredged.

Any gullies that form in the levees must be filled. Woody plants on the levee must be controlled by mowing, as their roots can create leaks in the levee, and attract burrowing animals that make more leaks.

Store Water in Open Tanks, Swimming Pools

Rather than roof and screen their water storage, some people prefer their cisterns *au naturel*. The big advantages to this approach are that you avoid the cost of a roof, and... you can swim in it!

Toss in some *Gambusia* (mosquito fish) to eat the mosquito larvae (but keep them out of natural waters). Either live with the algae or put some vascular plants in the system to pump the nutrients out of solution (they out-compete algae for nutrients). If you've got flow, you can use it to create a "skimmer" effect to clean off leaves and dust.

Natural pools, which are filtered and purified biologically instead of with chlorine, are possible but beyond the scope of this book, as they're not primarily for water storage.[15] If you want to be able to dunk without all these complications, you can put a *removable* lid on a small water tank. I've done this with a 900 gal redwood tank, and it worked great (photo, inside back cover). Aboveground swimming pools are the cheapest, funkiest storage going. Not a longterm solution, but you can't beat the cost. (See Really Cheap Storage, p. 50.)

It is an intriguing design challenge to create a more or less standard concrete swimming pool with functional cover, filtration, and skimming, to be usable both for storage and swimming—all at widely varying water levels. It appears that a level change of perhaps 14" *(35 cm)* is possible within these constraints. The idea is that the pool can be topped off with rainwater during the rainy season, then the level allowed to fall as water evaporates during the dry season so water doesn't have to be added to the pool. At a minimum, the pool should have built-in steps, and be designed so it is not a drowning or falling hazard.

Unlike a regular swimming pool, a "pebble tech" pool (with exposed aggregate on the inside) won't crack if left dry. (If you figure out more about how—or how not—to do this, please let me know.)

Store Water in Tanks

Tanks are the most common way to store water. A well-designed tank offers nearly complete control of storage conditions, including:

- security against leakage
- protection from mosquitoes and vermin
- shade so algae will not grow
- minimal or no evaporation
- valve-controlled inlets and outlets

We'll now spend the next three chapters looking at water tanks in detail.

Chapter 3: Water Tank Design

As we zoom in to focus on the design of tanks, and then plumbing details, remember to look up once in a while at the global view of your system and its context (Chapter 1). In this chapter, we'll be looking at:

- an overview of tank components
- situating water tanks
- sizing water tanks
- tank shape
- tank materials
- tank footings and floors
- tank roofs
- tank costs
- regulatory requirements
- hazards of stored water
- water tanks for special applications

Tank Components Overview

Most tanks will have:

- an inlet
- an outlet
- service access
- a drain
- an overflow mechanism
- critter-proofing
- air venting
- provision for sunscreen

We'll look at these components in detail in Chapter 4. Water tanks also can have a host of optional features, which we explore in Chapter 5.

All these valves, fittings, and doodads together add up to quite a list. When I design a water system, I **list every significant component in a table.** For each component, the table describes:

- function in the system
- generic name
- size
- material
- states in different system modes (dry season, wet season, fire emergency, supply interruption, and maintenance)
- how to replace it and what to replace it with when it fails

For example:

Component	Type	Size	Material	State in system modes	Replace with
Tank drain valve	Ball valve	2"	Brass	Open for tank cleaning after tank is drained as far as possible through outlet.	On onset of leaking, remove reduction and replace with 3" brass ball valve.

It's better to get a handle on the system's complexity and to work out design issues on paper—rather than with a saw or jackhammer. You can see more of this table on p. 84.

Situating Water Tanks

The location of your water tank will largely determine:

- which parts of your land can be supplied with tank water by gravity
- the amount of pressure at every point in the system
- the length and cost of pipe runs, control wire runs, and line-of-sight for radio links
- how visually obtrusive your tank will be
- the vulnerability of the tank and pipes to hazards such as falling trees, rocks, and landslides
- the size of tank it is feasible to build
- ease of construction and service access

McMansion in California with water tanks foolishly positioned below all uses. Water security would be far better with the tanks hidden in the middle of the house compound, at floor level or a bit higher.

To situate a water tank, you need to consider:

- elevation
- stability of soil and slope
- aesthetics, sacred spots
- security

Elevation

If you have a hill, put the tank at an elevation on it that yields adequate pressure. In places where there is no hill handy, you can:

- make a water tower (to artificially increase the elevation)
- use a small pressure tank (to pressurize water as it is needed—and have no water when the power goes out)
- use a huge pressure tank (to store pressurized water at low elevation)
- put a tank on your roof (and live with low pressure, like most people worldwide)

Pressurized from roof height, your appliances will barely work. With your tank the equivalent of ten stories above you, your washing machine, reverse osmosis water purifier, and demand water heater will start to work. With your storage 23 stories up, your fire hose will work optimally. With a tank higher than 23 stories, things will start to blow up.

Classic steel water tower in California's Central Valley.

The maximum advisable pressure for conventional plumbing is 100 psi.* You can always install a regulator to lower the pressure to whatever value you want.

If the water can make it into your tank by gravity flow, the only reason to limit its elevation is to have a shorter pipe run from storage to use, which can improve water security and lower cost.

Every home pressurized from a 260 gal (1 m^3) rooftop storage tank, in Manzanillo, Mexico. These provide water security, as the community system may only be on an hour a day.

For hydroelectric power, there is no maximum pressure—the more the better, period. However, most hydroelectric systems should be plumbed directly to the source, with no tank (except possibly a settling tank) in the line. If you are so lucky as to have a hydroelectric source high, high above your home, you can put the hydroelectric generator before your tank, so that you extract the extra energy before storing the water well above your home. If the flow is sufficient and the pressure is not, an option is to have the hydroelectric outlet be the headwaters of a fountain just below your home.

**Each 2.3′ of elevation adds 1 psi of pressure (each meter of elevation adds 10 kPa). Thus, 230′ of elevation difference produces the maximum recommended pressure of 100 psi (70 m produces the max of 700 kPa).*

What is the minimum elevation for your water tank? In terms of resource conservation, the lower the pressure the better. Outside of industrialized nations, very low water pressure is the norm; it works fine. If the only thing you need lots of pressure for is fire safety (possibly a legal requirement), it may be economical to install a booster pump or a separate, higher tank just for firefighting. **The higher you have to pump water to your tank, the higher your lifetime electric bill and environmental impact will be.**

Table 4 (below) shows the minimum and maximum pressures for several applications:

Tubs and kitchen sinks need flow, not pressure. If your pressure is very low, you can use larger-diameter pipe to get acceptable flow. (Figure 31, p. 87, shows a kitchen sink with almost zero pressure and lots of flow.)

A huge pressure tank for a water system that serves about a hundred homes. This installation is unusual in that all the storage is at the very lowest point of the system. The highest home served by the system is about 100' (30m) higher, and there is a perfect hill that's a 100' higher than that across the street. I guess they put the storage at the bottom because the water company doesn't own land anywhere besides where the well is. In case of a power outage, the pressure contained in the tank will push several thousand gallons of water into the distribution system before it runs out.

TABLE 4: WATER PRESSURES FOR DIFFERENT APPLICATIONS

Application	Min	Optimum	Max	Comment
	Pressure psi/ *kPa*			
Fire hose	40/ *276*	100/ *689*	100/ *689*	Excessive pressure can burst the hose
Drip irrigation	15/ *103*	25/ *172*	30/ *207*	Excessive pressure can burst the couplings
Hydroelectric		More=better	None	The more the better
Shower	5/ *34*	50/ *345*	100/ *689*	With a special showerhead you can use as little as an inch of pressure. Some people like the feeling of high pressure
Washing machine, dishwasher	10–15/ *69-103*	50/ *345*	100/ *689*	Pressure is needed to operate the inlet valves
Toilet	1/ *7*	Non-critical	100/ *689*	At very low pressure refill time is long
Tub	0.5/ *3*	Non-critical	100/ *689*	
R/O filter-standard	60/ *414*	80/ *552*	100/ *689*	
R/O filter-high flow	30/ *207*	40/ *276*	100/ *689*	Use high flow filter regardless—it lasts longer
Kitchen sink	0.5/ *3*		100/ *689*	

For conversion factors between height and pressure, see Appendix A: Measurements and Conversions, p. 90.

Elevated Water Storage and Structural Safety

One of the major consequences of deciding to locate water storage in a tower or on the roof is that it places a severe constraint on the amount of water that you can feasibly store.

A plastic tank which two people could roll onto a roof, empty, can hold enough weight of water to flatten the house. You will want to have any water tower or rooftop water storage engineered.

NATIONAL GEOPHYSICAL DATA CENTER

Engineer your raised tank well and it won't suffer the fate of this earthquake-destroyed tank tower in Iran.

Stability of Soil and Slope

You don't want your tank to sink into the ground, or slide down the hill. The load *per unit of area* from water tanks is actually quite low.* A person walking can easily place much higher point loads on the soil. On the other hand, no one has feet as big as a water tank. It's the *aggregate* load from a water tank—all that area being pushed on at the same time—that can push your building pad down into the gully.

However, undisturbed native soil is sufficiently strong to support even large tanks. In the case of a tank on a slope, where you don't have a natural flat spot, put the tank on cut (newly exposed, undisturbed soil), rather than fill (freshly dumped, loose soil). For a really large tank or any tank on fill, it's a good idea to consult an engineer. (See also Tank Footings and Floors, p. 48.)

> **Bad Location**
>
> *Some folks a few canyons over from us built a big (100,000 gal/380 m³) ferrocement tank on a steep, unstable slope. The whole thing slid several feet down the hill, and cracked open.*

Aesthetics, Sacred Spots

Water tanks can be big, and although they can be beautiful, they are more often ugly. When locating a water tank, either:

- put it where it doesn't matter
- conceal it well
- make it beautiful

Some places are so special that they just shouldn't be built on at all. This is often true of hilltops. Of course, the same topography that gives you a sunset view of all creation is also advantageous for cell phone and TV transmitters, and (to only a slightly lesser degree) water tanks. Unlike transmitters, however, you may be able to move the water tank down from the hilltop a bit to leave the silhouette unchanged, without compromising the functionality. You may even lower your pumping bill.

If you do make a tank on the tiptop of a hill, bury it part or all of the way, make it beautiful, ring it with trees, and/or make it easy to get on top of to hang out and admire the view.

Security

Ideally, you want your tank downstream from whatever hazards and weak links lie between you and your water source. Rivers that flood, gullies that wash out, landslides, falling trees, and rolling boulders—it's best if as few of these hazards as possible are between you and your tank.

Buried Storage

Burying your water tank has significant advantages over surface storage:

- less obtrusive
- cooler
- totally sunscreened
- more secure against accidental drainage
- considerable frost protection

...and some real disadvantages:

- except on a steep hill, you can't install a gravity drain, so the tank is difficult to clean
- usually requires a pump to get the water out
- design is more structurally challenging
- limits choice of materials to those that don't corrode
- can be a safety hazard
- water from the surface or surrounding soil can leak in and contaminate the tank
- inspection, repair, and replacement are more challenging

**An 8' deep tank might press on the earth 4 psi (28 kPa).*

Plastic septic tanks are false economy. Besides the inherent flimsiness, the one below, from Norwesco, had holes welded right through it at the factory. It cost the owners $1,000 to take this brand new tank out and replace it.

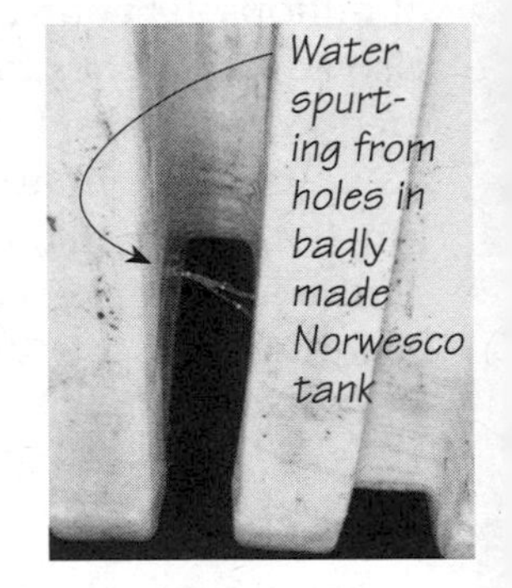

In my opinion, the disadvantages of buried tanks outweigh the advantages for most applications other than septic tanks. If stored underground, water from a source higher than the use point often winds up having to be pumped—to a use point downhill from the source. Pumping water downhill is surprisingly common—one of my pet design peeves.

Tanks designed for aboveground use generally shouldn't be buried. They may collapse under pressure from the surrounding soil after a while. While all tanks are designed to resist outward water pressure when full, tanks for burial need special shape and construction to resist *inward* pressure when empty. Such pressure can be considerable, especially if the soil around the tank is wet, contains expansive clay, is subject to frost heave, or is driven on by big trucks. To resist these forces, tanks designed for underground use generally:

- are deeply ribbed shapes or spheres, if made of plastic
- have thick walls of concrete (any shape)
- have high-strength walls of fiberglass in a cigar shape (like a big propane tank)

If you expect your buried tank to be empty with any regularity or for any length of time, it should be designed to resist this inward loading (spherical polyethylene tanks are).

Note that deeply ribbed polyethylene septic tanks, though designed for burial, are designed to **always be full of water** in order to push back against the soil that is trying to collapse the tank. When you pump such a septic tank, you are supposed to refill it immediately. This style of tank often collapses inward *despite* being filled with water. These are essentially cheap, disposable tanks, installed in a location (underground) where installation is expensive and disposal or replacement is difficult—a bad combination.

Wet soil can also pop a tank to the surface like a cork. Tanks are highly buoyant when empty. Even something as small as a 55 gal drum has 400 lbs of upward force on it when empty in wet soil. For a 1,000 gal tank, the upward force equals 4 tons—in the right conditions, enough to pop up through the asphalt of a driveway and lift a small car.[m]

Sizing Water Tanks

The size of your storage is one of the main factors that will determine under what circumstances you will find yourself short of water, and for how long. Will demand outstrip supply every morning? When there is a fire? A day after the well pump goes out? It will also do a lot to determine what your system costs.

Sizing your tank is a matter of figuring out what degree of water security you want, then finding the tank volume that makes the most of your water supply within your budget and other limiting factors. This is a good time to remember the reason(s) you want storage, as they will drive the calculation of tank size:

- You want more water security than a direct connection to the source can provide.
- The yield of the source cannot directly provide for peak demand.
- The yield of the source is less than that required for firefighting.
- The source is less secure than water stored in a tank (e.g., if the source requires pumping water, while water stored in a tank doesn't).
- The pipeline distance to the source is so far that it is more economical to use a smaller size pipe and a tank than a pipe large enough to carry the peak flow all the way from the source to the users.[16]

The biggest variable by far is how much water security you're aiming for. In general, the more storage you have, the better your water security. (See Perfection and Security Standard, p. 4, for a discussion of how water security standards tend to get overinflated.)

[m]**Metric:** *Upward lift on empty tanks in wet soil: 180 kg for 200 L drum, 3,600 kg for 4 m^3 tank.*

Without storage, the security—the percentage of time you've got water—is equal to the security of the source. The more storage you've got, the longer an interruption to the source supply you can cover with stored water (see Figure 9, below).

FIGURE 9: WATER SECURITY VS. TANK SIZE FOR DIFFERENT SOURCES

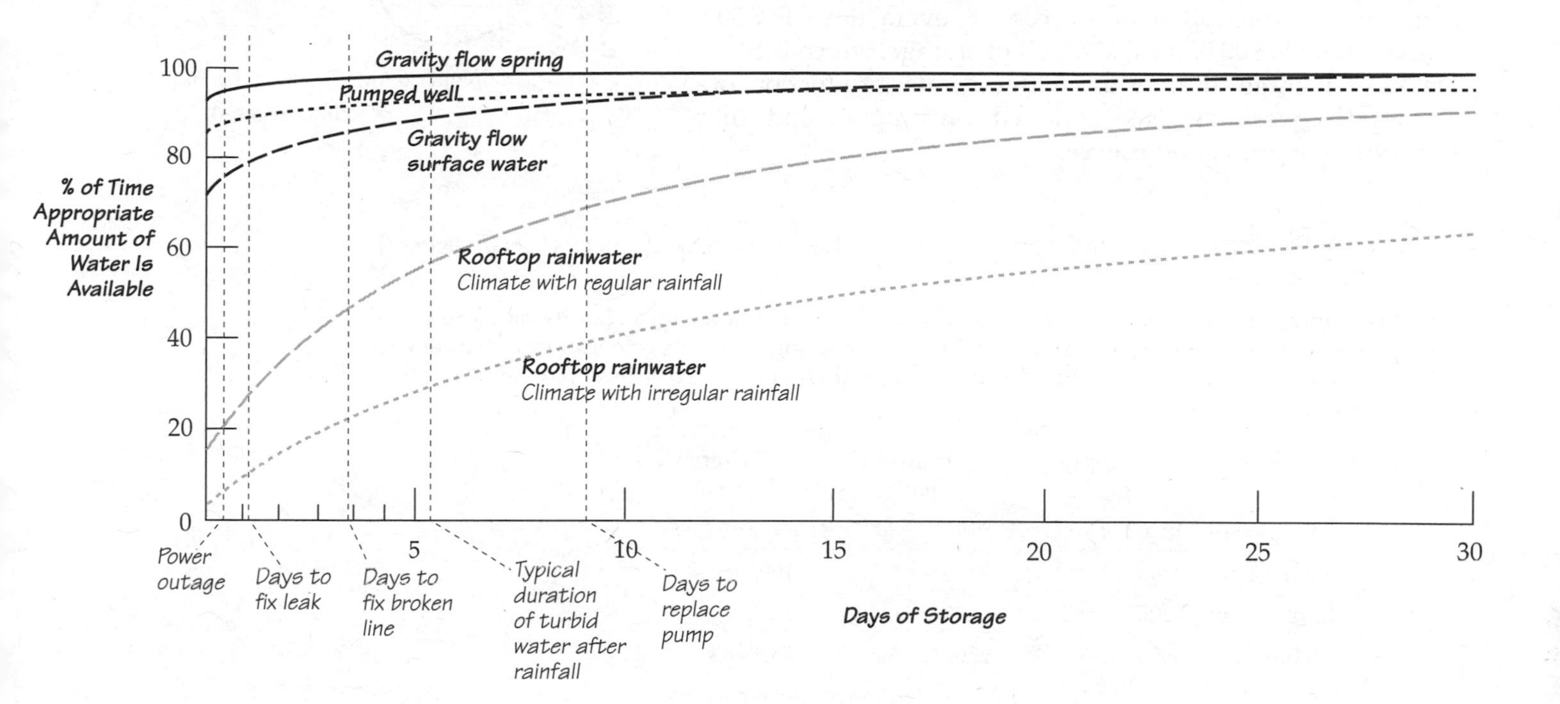

Is it possible to have too much storage? Yes. Too much storage can lead to freezing or water age problems (see Water Age, p. 12). More likely, it simply constitutes a waste of the Earth's valuable resources. Because of the high upfront cost of storage, it is rare to see anyone except the super-wealthy install too much storage volume.

There are factors that can lower the optimum amount of storage:

- **Not enough money**—Due to the high upfront cost of storage, you may wish to live with less than the optimum amount of storage initially or permanently.
- **Not enough space.**
- **Not enough elevation difference**—If the water source isn't much higher than the use location, you may not want to have a big tank use up a lot of the elevation difference (and pressure) between the tank's inlet and outlet.
- **Problematic access**—For instance, you may reduce your tank size if you have to hand-carry the tank or materials to the site.
- **Avoiding waste**—Even if you can afford the money and space, why waste the natural resources if more storage doesn't confer much advantage? (See Tunnel Vision, p. 5.)

If you hated word problems in math, you may wish to skip ahead to Tank Shape, p. 36...

Sizing a Tank for Demand Peaks which Exceed Flow

Although water needs are usually expressed as a value-per-person for a 24-hour day, in actuality just about all of this water will be used during a period of 10–12 hours. Over half of the entire day's water use may happen between dinner time and bedtime, or in the morning, depending on the culture. Water provided by the source during low-demand periods (e.g., overnight) can be stored for use during high-demand periods.

The minimum amount of storage that will not leave you short of water every day when usage peaks can be determined by making a table comparing the water coming into the storage compared to the water going out, throughout the day.

If extraordinary demand (e.g., for irrigation or weekend workshops) pushes consumption beyond the daily production for days at a time, you may need to look at a longer interval. One community, for example, has spells of extremely hot and dry weather, which last up to two weeks. Due to increased irrigation, the peak water demand can exceed the maximum combined production of all sources for several days. For 30 houses there is 100,000 gal *(380 m³)* of storage, which is two weeks of average summer use with no water production, or enough to cover two weeks of deficit consumption and still provide a generous fire reserve.

FIGURE 10: HOURLY WATER USE[17]

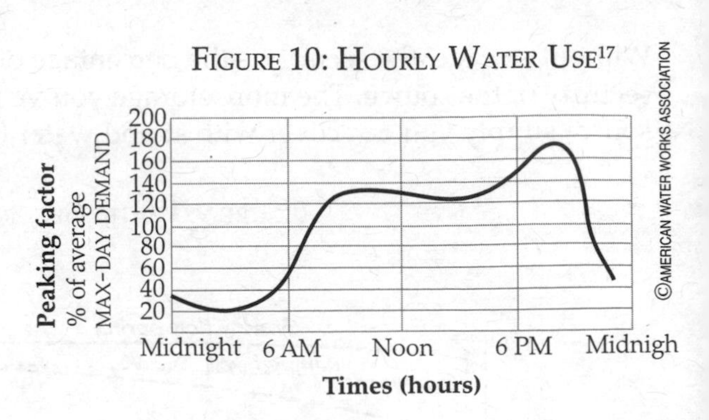

TABLE 5: SIZING A TANK FOR REGULAR, PERIODIC DEMAND PEAKS (SAMPLE CALCULATION)

In this example, the use for the system is 26,000 gal *(100 m³)* per day, with an hourly profile as in the graph above right. The supply is a 20 gpm *(115 m³/day)* spring. The table below shows the difference between how much water is produced and consumed throughout the day, most particularly in the critical evening hours:

	Supply gal/L	Demand gal/L	Difference gal/L
9pm-6 am	10800/ *41,000*	5100/ *19,300*	+5700/ *21,600*
6 am-12 pm	7200/ *27,000*	7300/ *28,000*	-100/ *380*
12 pm-6 pm	7200/ *27,000*	8400/ *32,000*	-1200/ *4,500*
6 pm-9pm	3600/ *13,600*	5200/ *19,700*	-1600/ *6,000*
		Largest deficiency	1600 gal/ *6 m³*

To provide for daily variation, a tank a bit larger than the largest deficiency would be indicated.

Note: The most common reason to size storage based on daily demand peaks is that you can't afford enough storage to cover emergencies or supply interruptions. If any of the sizing factors below come into play, your system will require more storage volume, and the daily demand peaks are moot. However, the same approach can be used to calculate demand peaks over any other time interval.

Sizing a Tank When You Have Limited Water Supply with Scheduled Use

This approach is the inverse of the approach above. It may be appropriate if the water supply is limited and there are known lengths of time without water use. Instead of sizing the tank to cover *use*, you size the tank to cover *production* of the source during the longest time *without* water use. If you store all the water that is produced during the longest time without usage, you'll have maximized your limited supply.

For instance, if water use occurs only during the day and the source goes 24 hours a day (like a spring), or has a limited 24-hour production (like a low-yield well), then the tank should be sized to hold at least the night's water production. This way, the full 24 hours' production is available during the hours of use.

A bottled water plant in Mexico plans to use water from a spring that yields only 2 gpm *(4 lpm)* in the dry season. If the plant is idle 16 hours a day, the full utility of the source could be captured with 2,000 gal *(7.6 m³)* of storage (16 hours of flow × 60 minutes × the flow per minute). If the plant is idle on Sunday, 32 hours of spring flow could be captured for Monday's production, using a 4,000 gal *(15 m³)* tank. Discretionary tasks that use a lot of water could be scheduled for Mondays to take advantage of the extra water.

Sizing a Tank to Cover Use During Interruptions in Supply

Most systems have at least a day's worth of storage to cover supply interruptions due to servicing, a fault within the system, or a disaster such as an earthquake or power outage. To size your tank for supply interruption, consider what is likely to jeopardize your supply and for how long (see Figure 9, p. 33). Also consider:

- **If your outlet is at the bottom of the tank, you may consume all your backup water before you figure out there's a problem**—You could install an alarm, but mid- and low-level outlets with valves will provide more security more simply. When your taps go dry at the mid-level, you can go to the tank and open the lower valve to access your emergency water. (You can also make a Variable Height Outlet, p. 68.)
  Warning: You don't want to have to go open a valve to get at fire reserve water. The preferred arrangement is to have the fire hydrant—only—connected from a lower-level outlet than the house/irrigation supply.
- **If you are aware that your water supply is interrupted, you can usually stretch your reserve quite a bit longer**—Through conservation.
- **The only storage big enough to cover year- or multi-year-long droughts**—Is in natural aquifers (see Conjunctive Use, p. 19), or large open reservoirs (a questionable technology—see dam discussion, p. 20).

TABLE 6: COVERING USE DURING INTERRUPTIONS IN SUPPLY (SAMPLE CALCULATION)

In this example, the average daily use for the system is 1,200 gal *(45 m^3)*. The only water source is a well. Experience has shown that the supply gets interrupted about this long, this often, and for these reasons:

Max days supply interruption	Average frequency	Cause
1	Once a year	Electrical blackout
2	Once in 2 years	Big leaks in supply line
1	Once in 2 years	Pull pump for scheduled maintenance
7	Once in 5 years	Unscheduled maintenance or replacement of pump

7 days × 1,200 gal = 8,400 gal of storage would be expected to cover supply interruptions that only occur once every five years. *(7 days × 4.5 m^3 = 31.5 m^3.)*
2 days × 1,200 gal = 2,400 gal of storage would be expected to leave you without water for a few days each five years. It would be easier to make up this shortfall by not irrigating for one week each five years and by using stringent indoor conservation, rather than by spending the extra money for a bigger tank. So, in this case, go with a 2,500 gal tank, the closest available size to 2,400. *(2 days × 4.5 m^3 = 9 m^3.)*

Sizing a Tank When Production Is Intermittent

If your water production is intermittent (for instance, from harvesting rainwater), your tank should cover the maximum cumulative deficit between production and consumption. There is a simple, graphical technique to calculate this:

- Plot a bar graph of average runoff from your roof by month.
- Plot cumulative runoff, by adding each monthly figure to all the previous ones.
- Draw a line equal to your cumulative water use.

Figure 11 (right) shows the final graph. The gap between the cumulative supply and cumulative demand shows the storage need or deficit. (See *Rainwater Harvesting and Runoff Management*[3] for more on sizing tanks for rainfall.)

FIGURE 11: GRAPHICAL CALCULATION OF STORAGE FOR RAINWATER HARVESTING[18]

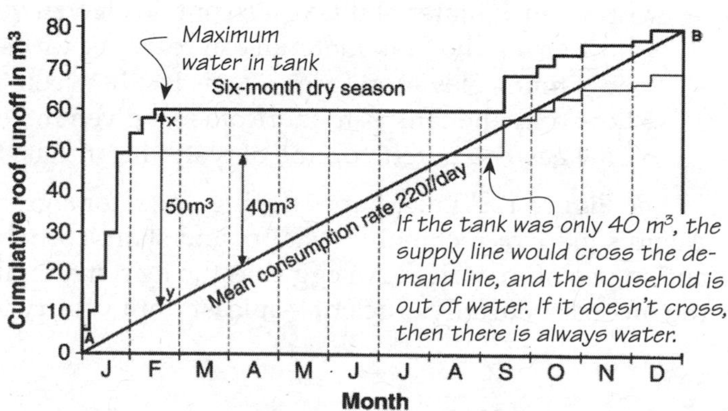

Sizing a Tank for Firefighting

If your system is part of a project that requires permits, there are likely specific legal requirements for the system's firefighting capabilities which you will have to research and fulfill. These probably include much more storage, bigger pipes, and higher pressure than you could otherwise imagine. A residence with its own water system may (for example) be required to have:

- a tank 20' or more from the structure (or fireproofed)
- 4,000 gal of storage, 2,000 gal of which is set aside for firefighting
- provision for the supply to automatically start refilling the tank when the level drops below 3500 gal

A subdivision might require:

- 6" hydrants with 4" and 2.5" outlets every 500'
- enough storage to supply 500 gpm at 100 psi for two hours from any of the hydrants[m]

There may be many pages of specifics about every aspect of the firefighting water supply in the applicable code; be sure to inform yourself about them early in the project. Ask your local Fire Marshal to review the plans for your system and provide suggestions.

See Systems for Firefighting, p. 78, for more on water for firefighting, including the use of foam to enhance the utility of a small amount of water.

Sizing Fire Storage

SSR = NFF + MDC – PC – ES – SS – FDS
—Or—
SSR = Storage Supply Required EQUALS
NFF = Needed Fire Flow PLUS
MDC = Maximum Daily Consumption LESS
PC = Production Capacity (based on the capacity of the treatment plant, the well capacity, or the pump capacity, depending on the system) LESS
ES = Emergency Supply (the water that can be brought into the system from connections with other systems) LESS
SS = Suction Supply (the supply that can be taken from nearby lakes and canals during the fire) LESS
FDS = Fire Department Supply (water that can be brought to the fire by trucks)

(All quantities are in flow units, i.e., volume per time.)

—Water Distribution System Handbook[19]

Size and Structural Integrity

As tanks get bigger, the structural engineering issues get much bigger. Tanks of a thousand gallons are no great challenge. A 10,000 gal *(40 m³)* tank requires serious consideration of the loads that will be operating on it. (See Appendix B: Tank Loads and Structural Considerations.) Any tank over 30,000 gal *(110 m³)* should be professionally engineered.

The tank shape determines how the material will resist the applied force and thus how easy it will be to resist a given load. This is something that you should consider carefully if you are making or modifying a tank. (See next section and Appendix B).

Tank Shape

Now that we've got the location and size of your tank, let's consider the shape. What difference does shape make? Shape affects:

- how much material it takes to contain a given volume of water (materials efficiency)
- how easily the tank material can resist the loads applied to it (structural efficiency)
- how much elevation (pressure) is lost between the top of the tank and the bottom
- how easy the tank is to fabricate in a given material
- how easily a given volume of water fits into its location

See Figure 12: Tank Shapes (facing page) for a graphical overview of tank shapes. Avoid square or rectangular shapes and sharp corners. These are inefficient structurally and in use of materials. An egg uses the least material to enclose the most water, and is the most structurally efficient—at least until you try to set it down on a flat surface.

---

[m]***Metric:*** *at least 6 m from structure, 15 m^3 of storage, 7.5 m^3 for firefighting, auto-refill below 13 m^3; 15 cm hydrants with 10 cm and 6 cm outlets every 150 m, enough storage to supply 2 m^3/min at 700 kPa for two hours.*

Figure 12: Tank Shapes

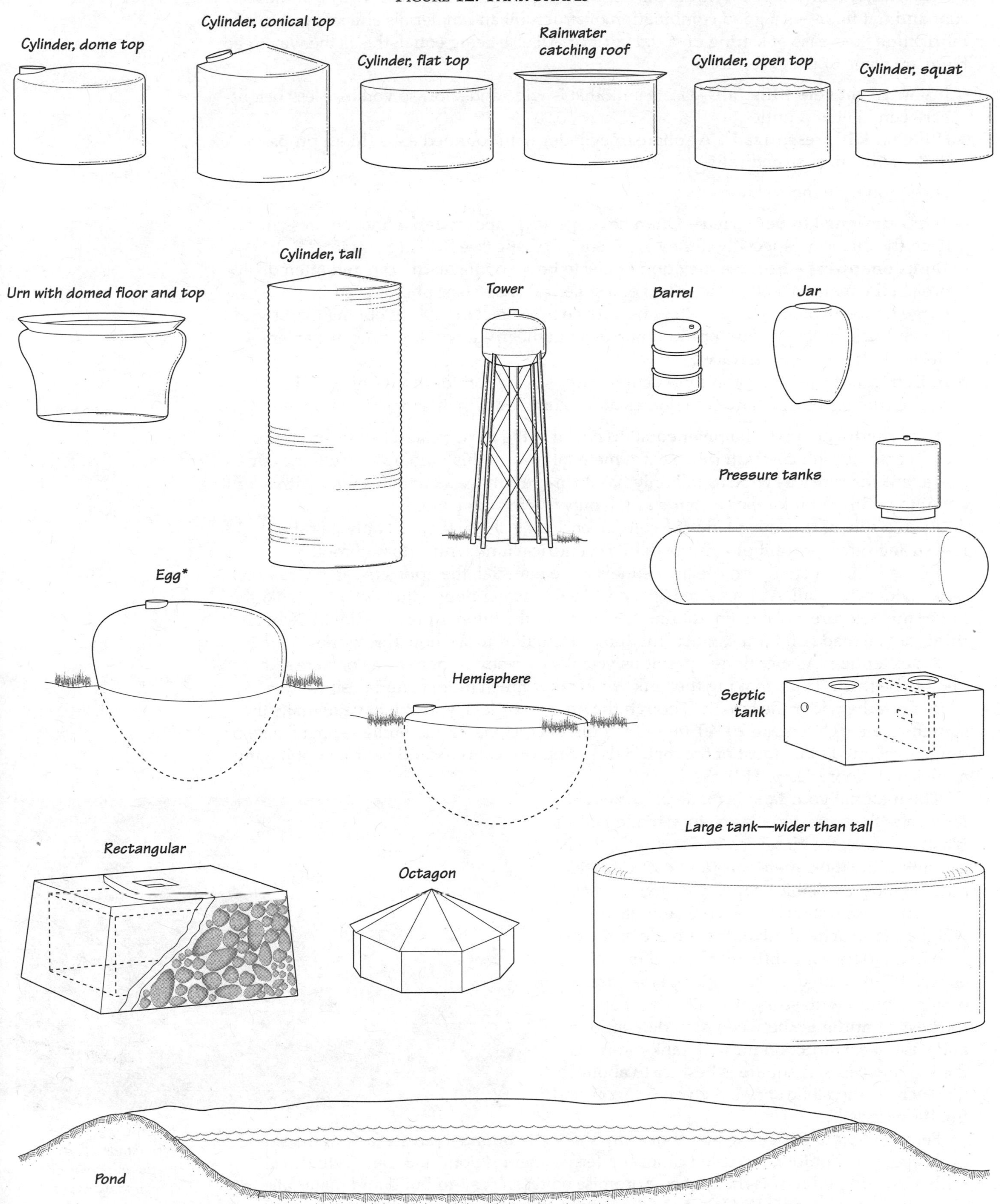

Never seen a 100,000 gal egg, but it would be way cool. This is the theoretically most material-efficient tank shape. Start with a sphere (lowest area to volume ratio). Squeeze the bottom so as water pressure rises, the diameter shrinks. This makes the hoop stress in the walls constant. The walls can be of uniform thickness then, like a clay jar. See p. 41, 91, 112 for more on materials efficiency.

The classic tank shape—cylindrical, about as big around as it is tall, with a domed roof and flat floor—is a good combination of structural and materials efficiency, ease of fabrication, and ease of setting on a flat surface. All else being equal, this is the way to go. Here are some of the exceptions:

- **If you've got very little fall**—Use a tank that is wide and short so you lose less height between inlet and outlet.
- **If the tank is pressurized**—A sphere or cylinder with rounded ends (like a propane tank) is the most structurally efficient.

Location influences shape:

- **Tanks designed to be buried**—Often have special shapes to resist uneven pressure from the outside, especially if they are made of plastic (see Buried Storage, p. 31).
- **Tanks on towers**—Because they don't need to be set on flat earth, can and often do approach the materials-efficiency ideal of a spherical shape (see photos p. iv, 29).
- **Large tanks on steep slopes**—May need to be made rectangular or oblong to fit (with the tank wider across the slope and narrower in the up-down direction, like an enclosed, water-holding terrace).
- **In tight quarters**—Tall cylinders occupy fewer square feet for more storage. The same can be true for buried tanks or squares and rectangles, if their shape fits in tighter.

For cylindrical tanks, diameter equal to height is the most materials-efficient ratio. However, it doesn't cost you much extra material to vary this ratio up to 2:1. For example, a tank 1.5 times as wide as tall only has 2% more surface area than a tank with height equal to width. A tank twice as wide as tall only has 5% more area.

Even a sphere (the most efficient volume-enclosing shape there is) only has about 13% less surface area per unit of volume than a cylindrical tank with a domed roof.

While it may seem that more area equals more material, the opposite can be true for tanks wider than tall. As pressure is proportional to water depth, the wider the tank, the lower the pressure for a given volume of water and the thinner the material the whole thing can be made of. (You can use our Tank Calculator[6] to see how this works.)

An exception: As mentioned previously, tanks with stiff floors, say a concrete slab, need to be thick. Thus, making the tank wider takes much more material, since there is more floor the wider the tank is. Though the pressure is less, you still have to make the slab thick enough to span 20–40' *(6–12 m)* without cracking. If you build a sphere, egg, or dome-bottomed tank, most of the materials savings in reality would be from not having a thick slab floor (See p. 112).

The material your tank is made of makes a difference with shape, too. For factory-made tanks of plastic or steel, the classic as-wide-as-tall cylindrical tank is the most materials- and cost-efficient, and the most available shape. For smaller steel tanks, the ease of fabricating conical roofs outweighs the structural advantages of domed roofs.

Circular tanks are difficult to build of rock, especially in small diameters. Rock tanks are easier to build with straight walls. The shape of the best compromise between materials-efficiency and ease of construction for rock tanks shifts as the size increases. A square is best up to about 10' *(3m)* across, then a hexagon, then an octagon, and finally, a circle.[16]

A 3,500 gal (13 m^3) ferrocement tank the shape and color of the surrounding rocks.

Ferrocement tanks can be just about any shape. If you must have a tank in the shape of an egg, urn, boulder, or curled dinosaur, ferrocement is your material. Cylindrical with a low-domed roof is the best compromise between easy-to-build and materials-efficient. (See Appendix D: How to Make Ferrocement Tanks.)

Tank Materials

There are plenty of choices for water tank materials, each with advantages and disadvantages. Table 7 (following page) summarizes their characteristics.

Materials Situations to Avoid

Despite all the contradictory data and opinions on the topic of how materials can contaminate stored water, a few circumstances are unequivocally hazardous and to be avoided:

- **PVC exposed to sunlight**—PVC breaks down in sunlight, reacting to form carcinogens, which leach into the water. It is a plumbing code violation to have potable water in unshaded PVC for this reason. You can see physical evidence of the change on the outside of the pipe; it darkens and becomes chalky and brittle. The reaction progresses from the outside in. To the extent PVC should be used at all, it should be buried or indoors. If you have PVC that has already degraded, you should replace it.
- **Pre-1997 PVC**—Which was made with more toxic plasticizers.
- **Flexible PVC water bed bladders or trash cans**—Which contain a high level of toxic plasticizers.
- **Pre-1980 tank coatings including coal tar and lead-based paint**—These were great for corrosion resistance but oops!—they poison the water.
- **Lead pipe and pre-1987 lead-soldered copper pipe**—Solders and flux currently contain less than 2% lead. Before 1987 they typically were half lead. Lead pipe can be recognized by its softness.
- **Western red cedar**—The same stuff that smells good and keeps it from rotting is toxic when ingested.
- **Fly ash in concrete**—Especially when exposed to acidic water.

Lead

Lead is a metal used for plumbing since Roman times. It is toxic if inhaled or swallowed. Just 15 parts per billion of lead is a hazard in drinking water (that's about five sand grains' worth per year). It accumulates in the body, and at high levels attacks the brain, kidneys, nervous system, and red blood cells.

If you are concerned about lead in your plumbing, have your water tested. If you can't replace the pipes, use the cold water tap for making tea and food. Hot water leaches more lead out of the plumbing. You can also run drinking water from the tap until it becomes colder before drinking it, using the flush water for houseplants or another purpose.

Often the worst hazards are not the base material, but solvents, additives, mold-release agents, fungicides, etc. to facilitate manufacture.

For more on leaching hazard by material, see the summary in Table 7: a Tank Materials, p. 40, and the material by material discussion which follows. There are also details on less common plastics in Appendix C, and another plethora of information in our Water Storage Extras download.[6] NSF International has a searchable database of products which meet NSF 61, their widely followed standard for materials in contact with drinking water.[20]

Glass

Glass is unequivocally the best material for storing drinking water. It imparts neither taste nor toxins, and can be washed, heated, and reused until broken, at which point it can be melted and recycled, endlessly, without degradation.

The weight and fragility of glass is an issue, but not an unmanageable one. In the village where I work in Mexico, the villagers are highly attuned to the taste and quality of drinking water. Five gallon *(20 L)* glass containers for transport and storage of drinking water remain popular despite the availability and greater convenience of plastic.

Toxicity/Leaching: No issue except with leaded glass.

Taste: Imparts no taste to water.

Table 7: Characteristics of Different Tank Materials

Key
y=*yes*, m=*maybe*
blank=*no or not applicable.*
CAPITAL, **Bold** = *emphatically so!*
1=*best*, 5=*worst*

Sizes available (darker = more common): 1, 5, 55, 300, 500, 2,500, 5,000, 50,000, 100,000, 500,000, 1,000,000+

Group	Material, construction	Life expectancy (in years)	Buy assembled?	Buy and assemble on site?	Make on site?	OK to bury?	Earthquake resistant?	Frost resistant?	Fire resistant?	Resists degradation by sunlight?	Protects water from sun?	Reusable, recyclable?	Material is biodegradable or innocuous	Ecologically benign	Can be relocated?	Suitable for pipes?	Suitable for transport of water?	Suitable for long-term emergency storage?	Inherently unlikely to leak?	Chemical hazard for drinking	Effect on taste	Comment
Masonry, glass, earth	**Aquifer**	Millions	N/A			na	y		y	y	y		Y	Y				Y	N	2	2	**Best choice for large quantities of water, esp. in arid lands**
	Glass	Until broken	y				n		y	y		y	m		y		m	y	y	1	1	**Best choice for drinking water**
	Ferrocement	40+			y	y	y		?	y	y							y	y	3	2	**Best permanent tank, DIY tank** (frost resistance unknown)
	Rock	30+			y	m[a]			y	y	y	y	y	y						3	2	Best for diversions; not tanks
	Pre-cast concrete	30+	y			y	y		y	y					y	y		m	y	3	2	Good for septic tanks, buried tanks
	Cast in place concrete	30+			y	y	m[a]		y	y								y	y[b]	3	2	Used for large, municipal tanks
	Unglazed, fired clay	10+	y		y				y	y	y	3	y	y	y	m[a]				2	2	Good primitive tech water storage
	Glazed clay	Until broken	y		y				y	y	y		y	y	y	m[a]		y	y	2	1	Good primitive tech water storage
	Brick and cement	30+			y	m[a]			y	y	y		m						m[b]	3	2	Good for small, custom size and shape tanks, valve boxes
	Unlined pond	Until it silts in			y		y		y	y	N		Y	Y					N	5	5	Good for large quantities of water, possibly wildlife, recreation
Steel	**Welded galvanized steel**	20+	y				m		y	y	y	y	y	y	y	y	y	m	y	2	2	**Good medium size tank,** large site-fabricated on-site tanks
	Bolted galvanized steel	30		y			m		y	y	y	y	y	y	y			m	y	2	2	**Good movable tank**
	Galvanized steel w/ plastic membrane	20+		y			m		y	y	y	y			y			m	y	3	2	Inexpensive
	Porcelain-bonded carbon steel	30+	y				m		y	y	y	m		y	y			n	y	1	1	Good for small tanks (rare)
	Epoxy-coated steel	30+		y			m		y	y	y	m		y	y			n	y	3	2	Popular for huge municipal tanks
	Stainless steel	100+	y				y		y	y	y	Y		m?	Y		y	y	y	1	1	Best for transporting water
Plastic	**High Density Polyethylene HDPE#2**	15, more in shade	y			m[c]	m			m[g]	m[g]	y		**m**	Y	Y	y	Y[d]	y	3	3	**Best cheap, easy to install, movable tank.** Used in "cloudy" milk and water jugs, water tanks, pipe
	Masonry in and over HDPE	100+			y				y	y	m								y	3	3	Could be the most durable tank of all. Sacrifices mobility for sun resistance and effective drainage
	GRP Glass-reinforced polyester	20	y				y[b]		N	m	m			N	Y		y	n	y	4	?	Excellent properties, but concern about toxins
	EDPM-lined pond	15+										N							m	4	?	**Best artificial pond liner**
	PVC (vinyl membrane)	10+		y					N			N	N	N					m	5	5	Avoid, due to plasticizers. Used in water beds, some soft bottles
	Aboveground swimming pool	10		y					N			N			Y				m	5	5	Low cost is the attraction
	Low Density Polyethylene LDPE #4	5	y									y		**m**	Y		y	y	y	3	4	Good for small, flexible water bottles and food storage bags
	PP Polypropylene #5	5	y								m	y		**m**	Y		y	y	y	3	2	Used in rigid containers, baby bottles, cups, bowls, bottle caps
	PETE #1 Polyethylene Terephthalate	5	y									y		**m**	Y		y	y	y	3	1	Used for mineral water
	Polycarbonate PC (cat. #7, "other")	5	y						N						y		y		y	3	1	Used in 5 gal water bottles, baby bottles, Nalgene® bottles
	PVC (rigid) #3 Polyvinyl Chloride	30+ (buried)		y			y		N[f]	N	m	N	N	N		m			y	4	1	Popular for pipe. Highly toxic in manufacture and disposal
Non-ferrous metals	Brass	50+	y				m		y	y	y	y		y	y	y[h]			y	2	1	Good for fittings
	Copper	50+	y		y		y		y	y	y	Y			y	y			y	2	1	Good for indoor supply plumbing, gutters
	Aluminum	30+	y				y		y	y	y	Y			y				y	4	2	Not great for water storage
Wood & leather	Redwood	50+		y	y					y	y	y	y	y	y				N	2	2	Beautiful
	Oak	15	y	y	y		y			y	y	y	y	y	y				N	3	3	Good for wine
	Cedar	20		y	y					y	y	y	y	y	y				N	4	5	Beautiful
	Leather	5	y		y		y			m	y		y	m	Y				N	3	5	Good for bota bags

Footnotes

a) Leaking is a concern for this use

b) If properly engineered and constructed

c) Ordinary plastic tanks can be buried 1–3′. Special plastic tanks are made for below grade use

d) Especially in thick-walled containers

e) See notes on p. 39 of hazards of PVC in sunlight or older PVC

f) Forms nasty toxins on burning

g) Offers sun protection if black, no protection if transparent. Black plastic lasts longer in the sun than clear

h) Really expensive

Ferrocement

Ferrocement tanks offer nearly the durability and strength of concrete at a fraction of the materials use, and with *complete* flexibility in shape. Ferrocement tanks are constructed from a grid of steel reinforcement covered with a sand/cement mix. The resulting wall is only two to four fingers thick, and is, particularly if curved, incredibly strong.

They are cost-competitive made to order, or you can make them yourself. Ferrocement is arguably the best all-around material for a permanent water tank.

The downside? You can't move them. And, you will *have* to make them yourself unless you live near one of the few folks who make them to order.[21, 22] They are labor-intensive and require some construction experience and ambition to build. We were unable to find information about the resistance of ferrocement to freeze/thaw cycles. (You can see the full magnitude of the do-it-yourself task in Appendix D: How to Make Ferrocement Tanks.)

Toxicity/Leaching: For best protection, use NSF 61 certified cement in the construction of your ferrocement tank, and NSF 61 certified sealers.[20] There is little concern about leaching from cement stucco after it has cured, which is mostly achieved within 30 days. The exception is with acid water or sulfides, which could continue to dissolve the tank.

Taste: May add a cement flavor until it is done curing, after which there is typically no detectable taste.

Galvanized Steel

Bolted or welded galvanized steel tanks offer high strength, medium durability, good fire resistance, and good transportability, and are overall an attractive choice. There is a substantial range in quality among galvanized steel tanks. The thicker the metal, the better. Corrugation generally indicates thinner metal. Welded steel is more common for small tanks. Large tanks can be welded or bolted.

Galvanizing works by letting zinc corrode in order to save the steel. The zinc gets consumed in the process. When it's used up, the steel underneath is left naked and unprotected and will corrode rapidly. If this is happening in just a few exterior spots, you could paint them with paint that contains zinc. However, I wouldn't trust this stuff on the inside of a tank containing potable water.

As the corrosion proceeds, it will result in lots of rusty powder, flakes, and sometimes, huge sheets sloughing off inside the tank (photo, right). If your tank is displaying this symptom, it would be a good idea to add an outlet screen. This will keep large pieces of rust from entering the plumbing and wreaking havoc there. There isn't much else you can do at this point except to save your pennies toward a new tank, or consider a repair membrane (p. 48).

Materials Efficiency

Tank size, shape, and material together determine the materials efficiency*—the ratio of the units of water volume to the units of material enclosing it. These ratios cover quite a range:*

- **2:1** for a rectangular, below-grade 1,000 gal *(3.4 m^3)* pre-cast concrete septic tank
- **12:1** for a lightest-duty 3,000 gal *(11 m^3)* cylindrical ferrocement tank with a domed roof and thin slab in non-industrialized nations
- **9:1** for the same tank with medium-duty ferrocement construction
- **6:1** for the same tank with heavy-duty ferrocement construction
- **140:1** for the same tank made of steel
- **70:1** for the same tank made of plastic
- **10:1** for a 100,000 gal *(380 m^3)* ferrocement tank made with flat slab and 3:1 width-to-height ratio
- **15:1** for a 100,000 gal ferrocement tank made with conical slab and 2:1 width-to-height ratio
- **25:1** (in theory) for a 100,000 gal, half-buried egg-shaped ferrocement tank (as yet untried, to our knowledge)

Materials efficiency is desirable in a world where the human footprint is ever-increasing. In general, more materials efficiency costs more labor and requires more attention to the design process. If you're interested in pursuing this topic, you might like to play with the Tank Calculator *in our* Water Storage Extras *download.*[6]

Giant sheets of rust peeling off the inside of an old galvanized bolted steel tank. Upper inlet has brass nipple in it; the lower inlet has a PVC male adapter. In each case the function is to keep the threads usable.

Accounts of short-lived galvanized tanks are generally traceable to mechanical damage to the galvanizing; uncontrolled water overflow over the outside of the tank, lack of a firm, free-draining gravel base; or use of electrolytically incompatible copper pipe or fittings. Properly installed, a good galvanized tank can last for many decades, and when it's finally done for, the steel is readily recyclable.

Toxicity/Leaching: Galvanized steel itself may leach both iron and zinc into the water, but neither of these are a great concern for humans. Zinc is highly toxic to fish, according to one source. If your tank has a liner, then the toxicity and taste issues will be determined by the liner.

Taste: Doesn't generally impart a taste unless rust is stirred up off the bottom, in which case the water will look and taste gross. But since iron oxide is neither particularly toxic nor soluble in water, rust is primarily an aesthetic issue.

A good quality, 50+ year old bolted steel galvanized tank (with smooth surface, in background) and a less than 15 year old cheap, riveted galvanized steel tank (with corrugations, in foreground). Though much newer, the cheaper tank shows signs that it will fail before the older, higher quality tank. The pile of parts in the foreground is for another good quality, bolted galvanized steel tank.

Bolted tank construction. The rubber gasket keeps the joints from leaking.

A really thick-walled, unplated coupling designed for welding into tank walls to form inlets and outlets.

Stainless Steel

The Cadillac material for liquid storage, stainless steel is so expensive it is almost never used for water, except for transporting it (see Tanks for Transporting Water, p. 54). Chlorinated water can corrode stainless steel.

Toxicity/Leaching/Taste: Stainless steel does not generally leach toxins, nor affect the taste of the water.

Porcelain-Bonded Carbon Steel

A rarity as a water tank material, baked enamel offers the strength of steel and the inertness of porcelain in contact with the water. I've heard that silo-rings of this material can be salvaged and adapted for use as water tanks.

Toxicity/Leaching: Generally low, however, some colors contain heavy metals.

Taste: No effect on taste.

Brass

Brass is good for plumbing, but rarely used for containers, due to its cost.

Toxicity/Leaching: Leaches copper and zinc, but does not pose a threat to humans.

Taste: No effect on taste.

Copper

Copper is rarely used for tanks due to its expense. Arguably the best material for rain gutters and indoor supply plumbing, it is common in these applications. Acidic water can deteriorate copper pipes and roofs.

Copper lasts a long time. The rate of leaching from copper decreases over time, presumably as the surface skins over with reaction products. Mining copper is environmentally devastating. Copper is readily recyclable.

Toxicity/Leaching: Copper leaches into water enough to kill microorganisms, but is of low toxicity to humans.

Taste: No effect on taste.

Aluminum

Aluminum is rarely used for tanks due to its expense, and rarely used for plumbing due to issues with electrolytic corrosion (the plumbing acting like a giant battery, with the metals electrochemically consumed). Absent acidic conditions or electrolytic corrosion, aluminum lasts a long time. Aluminum mining is environmentally damaging. Refining aluminum from ore takes prodigious amounts of electricity. Aluminum is readily recyclable.

Toxicity/Leaching: The rate of leaching decreases rapidly as the surface skins over with reaction products. However, acidic water can aggressively dissolve aluminum. Healthy people typically have low levels of aluminum because the digestive tract, skin, and lungs are effective barriers to absorption, and the kidneys efficiently eliminate absorbed aluminum. Although some studies have suggested a tentative link between aluminum and Alzheimer's disease and dementia, the evidence as a whole does not support a causal association.[23]

Taste: May add a slight metallic taste.

Rock and Mortar

It is tough to surpass a well-crafted rock tank for beauty and durability. However, the popularity of rock tanks as a choice for new construction of water storage is fading to zero in industrialized countries. This is a consequence of the large amount of skilled labor needed to build a tank, the staggering amount of material to build even a small tank, and their tendency to leak.

They remain an attractive option where labor is cheap, rock more accessible than bought tanks, and a bit of leakage to nourish their attractive patina of moss and plants no great loss.

I specify rock and mortar for diversions in natural watercourses. It just feels right, and is less intrusive than concrete and rebar. One real factor is that when floodwaters smash the diversion, it's just more boulders and sand in the creek bed, not an evil tangle of broken concrete and dangerously protruding, twisted rebar.

Dry-laid rock can be used for sunscreen and visual upgrade on tanks of other materials. (See Masonry in and over Plastic, p. 47, and photo at right.)

Toxicity/Leaching/Taste: Same as for ferrocement (they are plastered inside, so the water contact surface is the same).

A 12,000 gal (44 m³) mortar and rock diversion at the base of a waterfall in Mexico.

5,000 gal (20 m³) leak-plagued mortared stone tank for a community water supply in Mexico.

Aesthetic shield of rock around a 13,000 gal (50 m³) ferrocement cistern.

Concrete

Concrete is especially popular for large, municipal tanks. Most concrete tanks are made with reusable forms, which enable the considerable effort and expense of making suitable forms to be amortized over many tanks. Concrete tanks offer durability approaching that of rock, but with much less material and leaks. Pre-cast or cast-in-place concrete is good where high strength is needed to meet external loads, in buried tanks, for example.

Toxicity/Leaching/Taste: Same as for ferrocement.

Large concrete water tank at a retirement community of 2,000 members in Oregon.

Brick

This is another old-fashioned technique, good for small, square tanks in non-industrialized nations. It is much easier and quicker to work with than rock.

I think this technique is under-utilized in modern construction in the US, where it is suitable for small-sized tanks with specialized shapes, valve boxes, clean-outs, etc.. The Mexican masons on US job sites know how to do these things, if only they were asked…

Toxicity/Leaching/Taste: Same as for ferrocement (they are plastered inside, so the water contact surface is the same).

Pulling reusable steel forms off of a new concrete tank in a Mexican village.

Valve box with locking metal lid, made of brick in Mexico.

A cast concrete rain-filled tank for refilling fire trucks in the Los Padres National Forest.

Clay

Clay is rarely used for bulk water storage outside of non-industrialized nations. It is heavy, brittle—and beautiful. (See photo, back cover. Minerals from evaporated water are the source of the patina on this old *tinaja*.) Clay is an excellent material for storing drinking water. Ceramic urns of 55 gal *(200 L)* capacity remain common in Burma.

Toxicity/Leaching: Generally of low concern. Glazes may contain heavy metals. Of particular concern is lead glaze on low-fired pots from Mexico.

Taste: Unglazed, low-fired pots can add an earth taste to the water, but they keep it cool due to slow evaporation.

A 1,000 gal (3.8 m^3) pre-cast concrete septic tank.

Wood

Wood tanks were a common means of storing water in the past. They are beautiful and ingenious. The wood expands when wet, and steel hoops contain the expansion so the plank joints seal tight against each other. Wood tanks are most commonly made of redwood, cedar, or oak. Oak tanks are used mostly for wine.

Wood tanks have lost popularity due to a shortage of old-growth trees to make big, thick, close-vertical-grained heartwood planks—plus high cost, the necessity of keeping the tanks generally full to avoid drying out, and the tendency to leak slightly. The number of new wooden tanks outside the wine and hot tub industries has dropped nearly to zero.

The main limitation of wood for water tanks is that if you let the water level drop, the boards will dry out. If you raise the water level back up slowly enough, the walls will swell again and seal, and hardly any water will leak out. But if you let the *floor* dry out, you may lose the tank. Your best chance in that case is to tighten the lower hoops, then put a sprinkler or mister inside the tank. If you're lucky, the boards will swell enough after several hours that it will start to hold water again.

In areas where old wooden tanks were common, you may be able to buy them for a fraction of the value of the wood in them. I got a 900 gal *(3.4 m^3)* redwood tank for fifty bucks. I put new hoops on it, and sealed the cracks with special goop for this application.[24] The floor eventually buckled where it had nearly rotted through. However, it still holds water. The slow drip goes onto an orange tree, which is happy to have it. (Photo, inside back cover.)

There's no way to get a drain sump in a wood tank, but you can put in a floor drain, and install the tank tilted so the floor slopes toward it. To install a floor drain, drill a hole for the drain pipe, then mill a round depression around it with a router, then install a bulkhead fitting in it. Attach to the outlet using a rubber coupling with hose clamps. This will enable you to unhook the drain to move the tank, and will reduce the chance that the plumbing will crack as the tank shifts around.

Two 70 year old redwood tanks, one leaking, one not. The water in this system is highly sulfurous and quickly corrodes galvanized tanks.

Toxicity/Leaching: No issues, except the oils that make cedar so rot-resistant are toxic to humans as well as to microbes; a cedar tank is not suitable for potable water storage.

Caution: Some folks make water tanks out of treated marine plywood. I think both the shape (big, flat, weak surfaces) and the toxic, non-durable material are ill advised for water tanks.

Taste: May give the water a slight, non-objectionable wood taste.

Plastic

Plastic tanks are low cost, lightweight, and impervious. They are a good choice for small- and medium-sized tanks for residences and farms. The downside is that they are not available in big sizes, they turn into difficult-to-recycle trash relatively rapidly, and the bought tanks generally have a problematic combined outlet/drain.

All of these drawbacks except for the small-size limitation can be overcome by the techniques described in Masonry in and over Plastic, p. 47, at the cost of sacrificing the possibility of relocating the tank.

Which plastics are best? The danger of leaching from plastics depends on many variables. It is both poorly understood and controversial. The American Plastics Council, for example, has a different take than Greenpeace. Greenpeace is calling for a worldwide ban on PVC, a move gaining traction in a number of countries. The figure from

Greenpeace (at right), modeled on the food pyramid, proposes a proportion of plastic use based on ecological impact.

Toxicity/Leaching/Taste: The bottom line for water tanks is to use HDPE, and use as little PVC as possible. (See Table 7 for overall ratings by specific plastic, Appendix C for less common plastics, and Water Storage Extras[6] for an exhaustive survey of what is known on this topic.)

SUGGESTED PROPORTIONS FOR PLASTIC USE

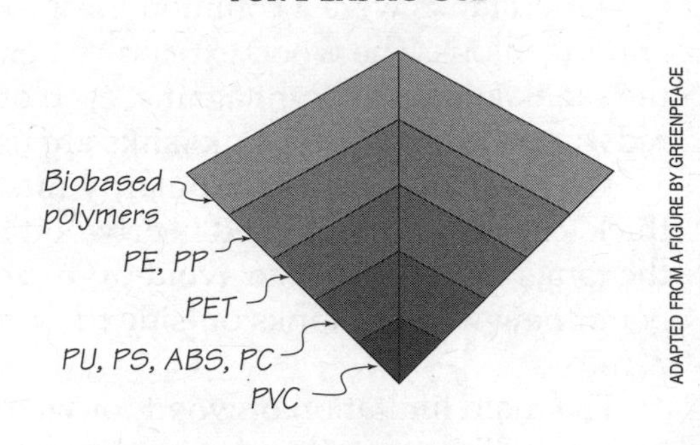

High Density Polyethylene (HDPE #2)

HDPE is the preferred plastic for water tanks, for which it is the most commonly used material. It is relatively innocuous in its manufacturing, use, and disposal, at least compared to other plastics. When the tank is no longer serviceable, the plastic can be reused. Polyethylene is not currently recycled—it is cascaded to uses with less stringent materials requirements.

High-density polyethylene is also the preferred plastic for plumbing. It can and should be directly substituted for PVC in most piping applications. However, the use of polyethylene for tank plumbing details is an alien practice in the US. One example of how such a connection can be accomplished is shown in Figure 18: Drain Options, p. 61. You can always insert a threaded adapter barb in a threaded in- or outlet and go with polyethylene from there. To reduce leakage in HDPE pipe, cover barbed connectors with silicone before clamping, or weld the lines.[16]

Toxicity/Leaching: Low toxicity.

Taste: Unfortunately, polyethylene can impart a plastic taste to the water, though this is rarely noticeable in large tanks. Keeping the tank shaded will reduce this effect.

Different height HDPE water tanks. Shorter tank is for drinking water; taller tank is for all other uses. The systems aren't connected at all.

Ethylene Propylene Diene Monomer (EPDM)

EPDM is the best artificial pond liner. It is a synthetic rubber resistant to heat, ozone, and UV light. It is able to stretch without tearing (there are photos of EPDM liners under Store Water In Ponds, p. 20).

There is little data on leaching of EPDM, although it is generally considered to be pretty inert. It is a more environmentally friendly building material than PVC.

EPDM is manufactured both for roofing and ponds. There is controversy about using EPDM roofing for ponds, with some claiming it can be used if washed. Actually, both EPDM roofing and pond materials need washing, as they are dusted with talcum powder to keep the plastic from sticking to itself. If storing potable water, make sure that the product meets NSF Standard 61 for contact with potable water.

In theory EPDM can be recycled, but it's not as easy as dropping it off at the recycling center. It is a thermoset material and cannot be re-melted. It can be ground up and used for something else. However, you may be hard-pressed to find a recycler to take an old pond liner off your hands.

Toxicity/Leaching: Low. EPDM sold as a roofing material may have a toxic coating.

Taste: Little or no effect on taste.

Plastic Taste

Waters bottled in PET plastic generally tasted better than those bottled in HDPE. That was true even within the same brand. [One brand], for example, was very good when bottled in PET, which imparted a hint of sweet, fruity plastic flavor (imagine the scent when you blow up a beach ball). But [the same brand] was only fair when bottled in HDPE, which made it taste a bit like melted plastic (imagine the smell when you get a plastic container too close to a flame).
—Consumer Reports

Fiberglass (Glass Fiber-Reinforced Polyester, GRP)

Fiberglass tanks are very strong, lightweight, and non-corrosive. Fiberglass is quite a bit stronger and more expensive than HDPE and generally considered to be higher quality. It is certainly superior to

HDPE for underground tanks due to its high strength. Exceptionally nasty solvents are used in the resin used to make fiberglass.

Toxicity/Leaching: There are reports of high concentrations of solvents in newly constructed fiberglass vessels (flushing is recommended). The literature is strangely quiet on the longterm health effects of drinking water from fully cured fiberglass tanks.[6]

Taste: There appears to be little effect on taste after full curing.

Epoxy-Coated Steel or Concrete

This is like a liner bonded to the tank. Epoxy-coated steel is a good choice for a large, durable tank. This is a popular option for big municipal-sized tanks. See Polyamide Epoxy in Appendix C for a look at the health and ecological effects.

Toxicity/Leaching: Moderate leaching concern. Make sure the coating meets the NSF 61 standard[20] for potable water. Also make sure that the epoxy is fully cured before filling the tank, and flush the first few tankfuls of water.

Taste: There appears to be little effect on taste after full curing.

Masonry in and over Plastic

You can turn a cheapo plastic (essentially disposable) tank into a first-class, high-performance, lifetime tank by combining it with masonry. This overcomes many of the shortcomings of both materials. In the conventional combination (metal/plastic liner), the metal can corrode and the light membrane puncture or tear. If you combine a plastic tank with ferrocement or stone masonry, there's nothing to corrode, and the thick plastic tank won't puncture or tear.

The full retrofit consists of:

- making a sloped masonry floor with a drain sump and dedicated drain
- adding a clean outlet
- adding masonry around the outside (see photo at right, and Plastic Tank Drain Retrofit, p. 60)

250 gal (1 m^3) HDPE drinking water tank with new bulkhead fitting outlet and chicken wire ready for stucco. (See also Plastic Tank Drain Retrofit, p. 60.)

Masonry outside provides complete shade and sunscreen, extending the life of the tank to approximately forever. The masonry can be dry-laid rock or stucco over chicken wire. The bigger the tank, the harder the retrofit gets.

A masonry floor retrofit provides better drainage and cleaner outlet water. The masonry should be added with the tank full of warm water, i.e., at its maximum expansion, so the plastic and not the masonry takes the tension load.

Toxicity/Leaching/Taste: Same as for a plastic tank.

Galvanized Steel with Plastic Membrane

The addition of an interior plastic membrane enables a much lighter-weight galvanized tank to have an acceptable life span. The trade-off is enjoying low cost versus losing the flexibility to add inlets and outlets, reducing repairability and maintainability, and increasing the difficulty of getting a good drain.

This composite tank design uses very lightweight steel for strength, and a membrane for waterproofing. It seems like a very efficient use of resources. However, I have a hard time getting myself to truly embrace something made out of plastic and the thinnest possible metal. It essentially amounts to repairing the tank in advance, knowing it is too thin to hold water for long. The brilliance of it is that without water on the metal, it will last longer, and installing the membrane from the get-go is cheaper than retrofitting it later when the tank is old, wet, and rusty. This is a relatively new tank technology. Time will tell how well these systems hold up.

50,000 gal (190 m^3) galvanized steel tank with plastic membrane liner.

Toxicity/Leaching/Taste: Leaching hazard and taste depend on the membrane material, which can be PVC (bad), epoxy (supposedly OK after curing), or polyurethane (bad in production, not clear how it is in use[6]).

Interior Membranes for Repair

If a structurally sound tank of any material starts to leak, you may be able to get some more life out of it by adding a plastic membrane inside.

Toxicity/Leaching/Taste: Depends on material, same as Galvanized Steel with Plastic Membrane, above).

Plastic Bladders

This is a class of storage where the water is contained within a non-structural plastic membrane, which is supported by some other structure. There are many options here, ranging from a waterbed bladder, to a system where a giant slug of a membrane is cocooned in underground culverts.[25]

I'm not particularly drawn to this approach. One objection is that plasticizers in flexible membranes tend to be more toxic than the same plastic in rigid form. (Don't drink from your waterbed—see Materials Situations to Avoid, p. 39.) Beyond that, it just seems sketchy to store a boatload of precious water in what amounts to a plastic bag. I would think it would pinhole all over, or possibly tear unless you were fanatically careful about the installation and service. However, in practice, these systems seem to be working so far, and they do offer tremendous cost savings.

Toxicity/Leaching/Taste: Depends on material, same as Galvanized Steel with Plastic Membrane, p. 47.

Goat Bladders, Leather, Etc.

Animal skins have long been used to transport water. Leather vessels keep the water cool by sweating through the walls.

Well into the dispiriting research on toxic leaching, I found an account from another researcher who was so appalled by the threats that he was ready to go back to storing water in goat bladders, like his ancestors. I have to admit that after reading countless abstracts of studies about nasty chemicals from plastics, contaminated cements, etc. the idea does have a certain appeal.

Toxicity/Leaching: In the case of leather, tanning chemicals may enter the water.

Taste: Most skins impart a taste.

Tank Footings and Floors

A firm, well-drained footing is essential for long tank life, especially for large tanks. These footing considerations apply to all tanks:

- The earth under the tank should be well-compacted and free of large or sharp rocks.
- The surface drainage should be away from the tank in all directions (except for some runoff harvesting tanks).
- Tanks on benches cut into a slope should be resting entirely on undisturbed soil (cut), not on tailings from the excavation (fill).

Steel tanks are usually set on a bed of gravel. This slows corrosion of the bottom of the tank by keeping it dry underneath. The compacted soil under the gravel is ideally sloped to one side (or all sides—see figure at right) so moisture is less likely to get under the tank, softening the soil and condensing on the tank bottom. Coarse gravel is preferred, to promote drying air circulation under the tank.

Plastic or fiberglass tanks can be set directly on firm, rock-free soil. If there are rocks, or the surface is uneven, the floor of the tank can be protected with a thin layer of compacted sand or pea gravel that won't wash away. (Plastic isn't strong enough to bridge the large gaps between coarse gravel.) For tanks which bear on the footing with over 800 lbs / ft² *(3900 kg/m³)*, a concrete footing with steel support stands is suggested to prevent movement in wind or earthquakes.

The floors of ferrocement or concrete tanks can be poured directly onto firm soil free of large rocks. If the natural surface is too uneven, it can be smoothed with the addition of a layer of sand.

The walls of a concrete tank should be buried at least to the level of the floor inside the tank. This is to reduce the chance that erosion will undercut the floor, possibly leading it to crack.

Large roots under a tank could heave, possibly cracking the floor. If the roots can't be removed, they can be covered over with a raised layer of gravel. Roots won't grow into the gravel if it doesn't have water in it. If the soil under the tank is disturbed, re-compact it well.

If the only available site has huge rocks or bedrock under it, try to build entirely on the rock, otherwise the difference between how the soil and rock compress could crack the tank.

A community in Mexico built a 100,000 gal *(380 m³)* cistern by sealing the space between vertical walls of exposed bedrock with a rock wall and pouring a footing on the earth between. This cistern has always leaked, and the community spent as much time and money trying to seal it as to build it in the first place. I suspect that the problem is that the earth under the tank moves slightly as 400 tons of water press down on it and then are removed, while the bedrock holds still.

If your footing is perfectly smooth and stable, the load on the tank floor will be insignificant. If it is not, the resulting strain can crack the floor and walls. (For an explanation of the radically different ways flexible and stiff tank floors of different shapes behave structurally, see p. 91, 112.)

Plastic, fiberglass, ferrocement, or concrete tanks can be partially buried. The tank should be capable of withstanding the press of wet soil from outside in without collapsing, and should be leak-free so water moving through the soil does not contaminate it.

I would hesitate to bury a mortared rock tank. There is a strong likelihood that roots will find some crevice in the masonry, work their way through it to the water inside, and possibly damage the tank as the roots grow in thickness.

Figure 13: Tank Footings

FOOTING FOR GALVANIZED WATER TANK

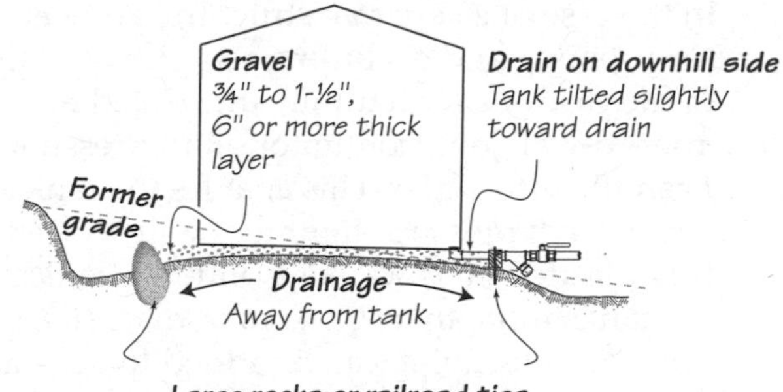

FOOTING FOR PLASTIC OR FIBERGLASS TANK

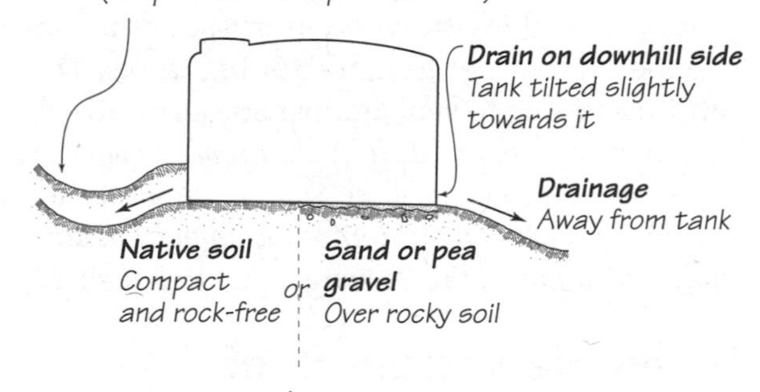

FOOTING FOR FERROCEMENT OR CONCRETE TANK

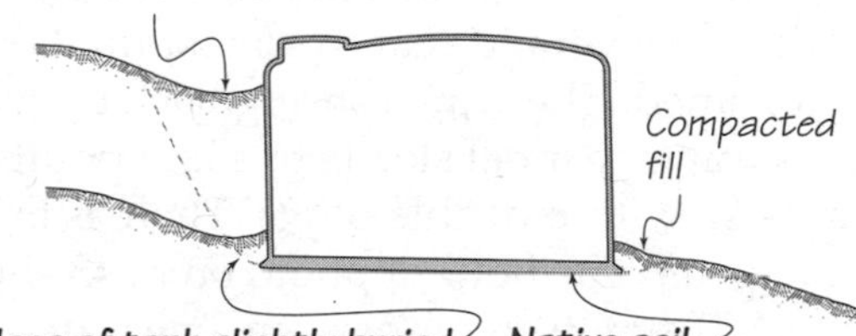

An old concrete slab for a redwood tank, retrofitted with railroad ties and gravel cover to receive a bolted galvanized steel tank (sheets at right). The tongue going into the far side of the gravel is to provide a clear path for the drain line.

Tank Roofs

Roofs can be structurally integrated with the walls, or a separate structure set on top.

In the case of a separate structure, good attention must be paid to details of critter-proofing and excluding roof runoff at the roof-to-wall interface.

In the case of a structurally integrated roof, it can be advantageous to make the roof-to-wall joint continuous and pressure-tight, so the space under the roof can fill with water. This enables the tank to store significantly more water. An integrated, pressure-tight roof will typically be of "unibody" material such as plastic or ferrocement, and in a conical or dome shape.

For tanks more than 15' or so in diameter, a central pillar of steel-reinforced concrete or a steel pipe can be added to help support the roof. This can be a convenient place to add an integral ladder.

Roof supports in a 50,000 gal (190 m³) bolted steel galvanized tank.

Domed roofs are most structurally efficient. However, if you intend to make the roof of concrete, wood, or steel, a flat roof will be much easier to build or support with straight materials like wooden beams. Conical, hexagonal, or octagonal roofs are compromises that offer advantages of both flat and domed roofs.

(See Appendix B for more information on the structural properties of different shapes.)

Roofs made with wooden trusses and conventional roofing (like a house roof) can be cheaper and easier to make for big tanks. They have the disadvantage that they are all but impossible to seal against spiders and whatnot.

Note for galvanized sheet steel roofs on square or rectangular tanks: It is easier to slightly adjust the dimensions of the tank so the sheets neatly cover it (e.g., "5 sheets wide by 1.5 sheets long"). This helps to minimize the amount of cutting, which is a relatively difficult task. For multi-sided tanks, this is more difficult, but should still be kept in mind.[16]

Water-Harvesting Roof

If your tank is for harvested rainwater, why not harvest the rainwater from the roof of the cistern as well? This is easiest with a cement tank into which rainwater harvesting "wings" can easily be incorporated—see large photo on front cover, top left photo on p. 57. The latter has 2' *(60 cm)* "wings" which increase catchment from its own roof about 40%. With 60" *(1.5 m)* of rainfall a year, this 13,000 gal *(50 m³)* tank will catch enough rainfall in an average year to almost fill it once—a significant contribution. The cistern roof is domed. The wings spring back up from the low point at the rim of the dome. The resulting channel slopes to a low point, where there is both a drain and an inlet into the cistern, with movable plugs. There is a photo of this cistern from below in Rock and Mortar, p. 43. The bottom photo on p. 43 shows a cistern which fills *entirely* from water harvested from its own roof—in a near desert at that. There is more on water harvesting roofs in our book *Rainwater Harvesting and Runoff Management*.[3]

Tank Costs

Figure 14 (right) summarizes typical costs for different size tanks. For all materials, cost per gallon drops steeply at first, then less dramatically as tank size increases.

Really Cheap Storage

If economy is an overriding consideration, here are some suggestions for really cheap storage:

- **Salvage 30 or 55 gal drums**—Can often be scrounged for little or nothing. The bungs often have ¾" pipe threads, facilitating attachment of inexpensive plumbing, for example, small diameter polyethylene tubing. Drums can also be drilled, tapped, threaded, or fitted with bulkhead fittings to make inlets and outlets in any position.

- ❖ **Tote bins**—Used for palletized, bulk transport of liquids. They can be made into passable small tanks, if they've contained something non-hazardous. They are usually 275 gal *(1 m³)* HDPE containers.
- ❖ **Aboveground swimming pools**—The cheapest, funkiest storage going. Not a longterm solution, but you can't beat the cost. The plastic walls are usually PVC.
- ❖ **Ponds**—Can be relatively inexpensive for large volumes of water.
- ❖ **Aquifers**—Usually don't cost anything and can store vast amounts of water.

Free, salvage 275 gal (1 m³) tote bin. It's roughly cubical and about chest-high.

ZODIAC POOLS 26

Suburban water storage technology transfers readily to backwoods pot growers and shoestring settlers.

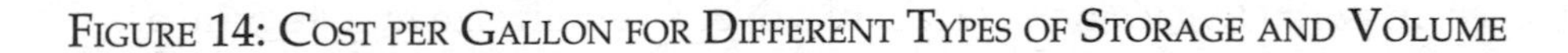

FIGURE 14: COST PER GALLON FOR DIFFERENT TYPES OF STORAGE AND VOLUME

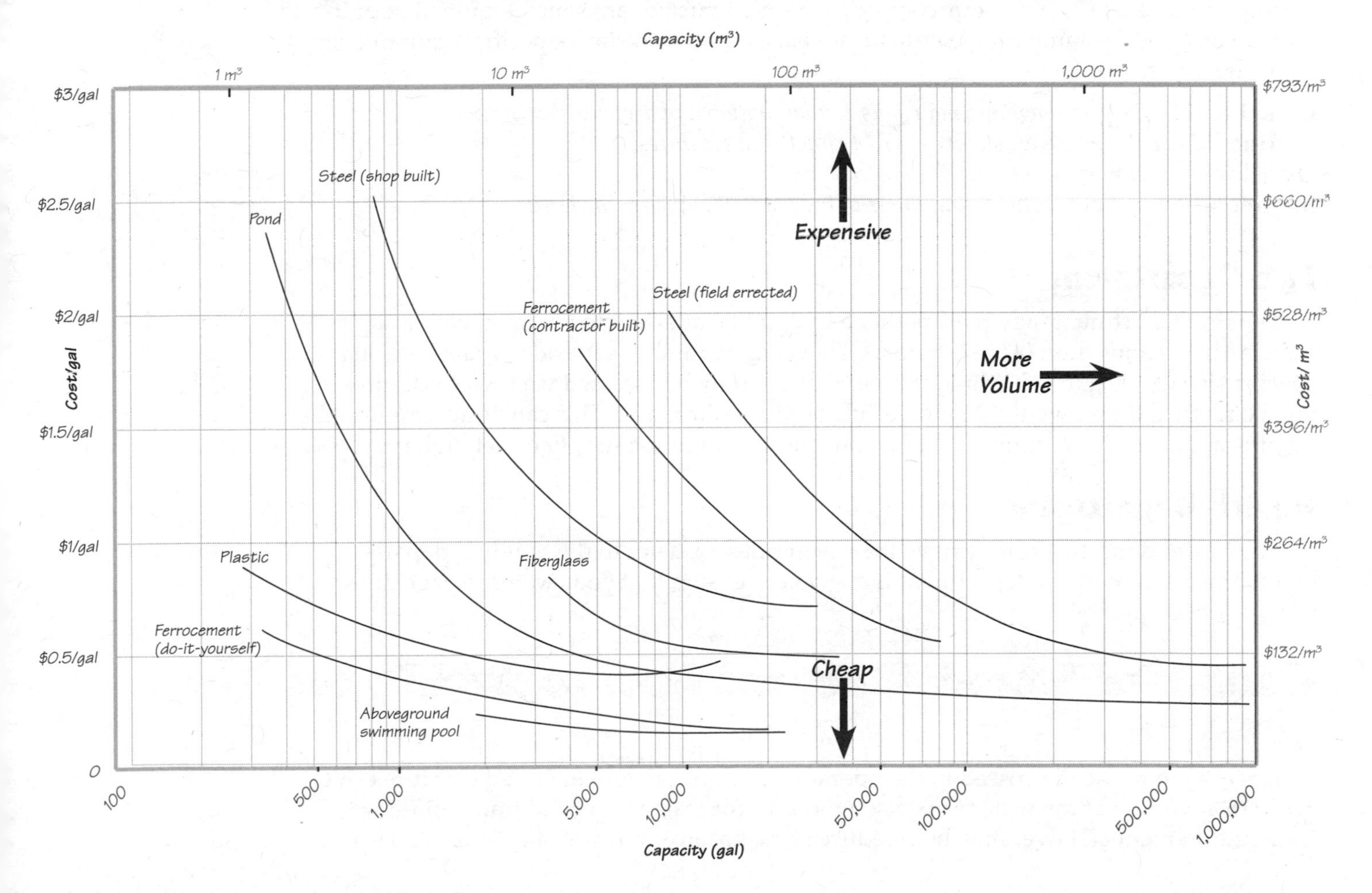

(2005 dollars; some costs have risen steeply)

Regulatory Requirements

Many, if not most, water tanks are installed with little or no regulator involvement, but rules and enforcement can vary quite a bit depending on where you are. You'll need to inquire locally to find out what you'll be subject to.

Water storage may be subject to zoning, building department, fire department, and health department rules. If you have a homeowners' association, it may regulate water tanks, perhaps just because they are a structure. Your insurance company may have rules or incentives relating to water storage as a fire safety resource, a flooding hazard, or simply as another asset to insure.

Some of the rules you may encounter will be consistent with your own interests, and some will run counter to them. You may run into rules concerning:

Zoning

You may or may not be allowed to build a water tank within the building setbacks from your property lines—the zoning department can tell you.

Architectural Guidelines

In some neighborhoods, rules may prohibit aboveground tanks. A beautiful ferrocement tank with an attractive shape and color, or one that looks just like a boulder, may be able to overcome anti-tank prejudice.

Building Department

It will generally be the building department that enforces plumbing code requirements about pipe sizes, materials, placement, etc. Tanks over a certain size may require a permit—5,000 gal/*18.9 m*^{3} in our county. Large, constructed tanks may require a permit with an engineer's stamp on the structural plans. Some of the tank-specific requirements we've heard of:

Lockable lid—*To guard against malicious contamination and drowning hazard.*
Sealing lid—*To guard against entry of roof runoff and creatures.*
Overflow—*With mosquito trap.*
Soil report—*For the structural soundness of the soil that supports the tank.*

Fire Department

The fire department may require that a hydrant be attached to the tank, with a certain-sized connection (4" in our area). They may require a "set-aside" or reserve of a specific size (2,000 gal/*9.5 m*^{3} in our area) that can only be accessed via their hydrant—not a bad idea, if you want them to be able to save your house. This can be accomplished by putting the hydrant outlet at the bottom and the domestic supply outlet higher up.

Health Department

The health department may have their own rules, or defer to the building department. The vector control department may want to ensure that your water storage does not breed mosquitoes.

Hazards of Stored Water and How to Avoid Them

It is easy to get so engrossed in the operation of the system as intended that it doesn't occur to anyone that the water tank also works as, for instance, a child trap. Fulfilling legal requirements (above) may help reduce some hazards, but you should also take a direct look at hazard reduction.

(Note: This section covers ways that stored water can harm things. Ways that things can harm stored water are covered under Protecting Stored Water, p. 77.)

Drowning

Drowning can be a real hazard with stored water, primarily for children. The two strategies to reduce drowning hazard are:

- **Limit access to the water,** e.g., with a fence, locked access hatch, or removable outside ladder.
- **Provide an easy way out of the water,** e.g., a built-in ladder on the inside of the tank, or steps on the sides of a reservoir.

A water tank with an access hatch in the middle and no built-in ladder is especially hazardous. Even a resourceful adult might drown before they could figure a way out of the tank. At least leave a knotted rope dangling down.

Structural Collapse

What will happen if the water tank topples? Could a water tower fall over and squash a house? Maybe it should be a bit farther away. Could an earthquake or landslide tip a tank off of its pad on a steep hillside, with the result that it steamrolls over houses below?

The engineering of water tanks is not intuitively obvious. The structural loading per unit of area is usually lower than people think, and the total loading higher than people think. (See also Appendix B, Structures, p. 91, and Protecting Stored Water, p. 77.)

Tank ruin in the field where I played as a kid. I don't know what did this tank in, but it makes good lizard habitat now.

Flooding

What will happen to the water if a tank or pipes break? If the answer is "not much," then you don't need to dwell on it. If a life-threatening situation might result (like the one described on p. 91), safety may be the main design factor.

Pestilence

Biological hazards in stored water should be minimal or nonexistent with good storage design and if the water was clean in the first place. (See How Water Quality Changes in Storage, p. 9.)

Toxic Contamination

Assuming there are no toxins in the incoming water, the route for toxins to get into stored water is by leaching from the storage vessel or plumbing. Toxins from outside can also permeate *through* the walls of plastic containers. Leaching is discussed in general in How Water Quality Changes in Storage. Leaching hazards from various materials are discussed in Table 7: Materials, p. 40, and in more depth in Appendix C. Permeation is discussed on p. 78. There is a bunch more info on this issue in Water Storage Extras.[6]

Liability Exposure

If your water system incorporates a hazard that you should have been aware of, or if your water storage doesn't meet legal standards, you could be sued if someone or something comes to harm as a result. Good design will minimize the chance of this happening, or your being liable if it does.

Water Tanks for Special Applications

Pressure Tanks

Pressure tanks are an alternative to an elevated tank to provide pressure. They typically have a pressure switch that controls the pump that pressurizes them, an air bladder that stores enough pressure to push out a third or so of the volume of the tank before the pump has to switch back on, and an air valve to adjust the air pressure. Pressure tanks are usually small—30–50 gal *(75–190 L)*. Their advantage is that they are far less expensive than an elevated tank and the piping to it.

Their disadvantage is that they provide little water security because when the power goes out, there is almost zero reserve. If you can, situate the main tank higher than the water use points, so that water will still flow, without power, albeit at low pressure.

As a last resort, you can put in a really big, expensive pressure tank to get more reserve (see photo, p. 30).

Pressure tank and pressure pump.

Break Pressure Tanks

These aren't for storage, though they do have "tank" in the name. Break pressure tanks provide an air gap that releases the contained pressure in a pipe. They then funnel the water into another run of pipe. Strategically placed, these tanks can reduce the amount of expensive, pressure-resistant pipe in a system. A break pressure tank has an inlet line, provision for venting and access, and an outlet. They are usually small, accommodating just a moment's flow.

Hot Water Storage

Some of the same techniques described under Freeze Protection (p. 73) can be used to make a tank that holds warm or hot water. Hot water tanks and some solar water heaters use the "thermos" effect to great advantage—a high vacuum between walls of a glass pressure vessel prevents heat loss by convection or conduction, and a silver coating reflects radiant heat loss.

Hot spring water is often highly corrosive, and especially resistant materials may be required for plumbing. One hot spring found CPVC to be the most resistant material; for another, only copper pipe with high-temperature silver solder would hold up.

Tanks for Transporting Water

For transport, the ideal water container would be strong and lightweight, would resist sunlight, and would neither leach nor slough off nasty stuff into the water. In a stationary tank, great sheets of rust that peel off just harmlessly settle to the bottom, but in a mobile tank, the water sloshes and stirs everything up. Stainless steel is ideal, followed by polyethylene.

Transport of water from a spring in Cuba.

Stainless steel water tank/bottle truck proposed for distribution of inexpensive, bulk drinking water refills for an Indian village in Mexico.[27]

Chapter 4: Common Features of Water Tanks

This section covers the features common to almost all water tanks. There is more information on optional features in Chapter 5. (Inlets in general are covered here, while you'll find Inlet Float Valves there.)

FIGURE 15: COMMON FEATURES OF WATER TANKS

Access
For service/inspection with critter-proof lid
Vent
Mosquito-screened
Inlet
Max storage level
Overflow
Storage to cover peaks in use
Check Valve
Outlet shut-off valve
Low level outlet
To access reserve
Firefighting reserve set-aside (optional)
Dead storage
Sloped floor
For ease of cleaning
Drain sump
Hydrant

Inlet

An inlet is where water flows into your tank. The preferred practice is for each water source to have a separate inlet. The type of water source determines where on the tank your inlet needs to be located.

If the water source is a well, or otherwise below the tank, the inlet must have an air gap to the surface of the water, and must be a pipe diameter above the highest level the water can reach when the overflow is flowing at full capacity. This is to avoid siphoning tank water in reverse through the inlet line, possibly contaminating the well (see Figure 19, p. 62). The inlet should be as high in the tank as possible so that the overflow can be high. The spill point (lowest point) of the overflow, in turn, determines the maximum useful storage capacity of the tank. A high inlet/overflow will help get your money's worth out of the tank capacity.

If the water source is gravity-flow water, the inlet can be just about anywhere. The tank water can't flow back uphill to contaminate the source. An inlet at the top is convenient for installation of an inlet float valve to automatically shut off the incoming water when the tank is full (see p. 66). An inlet near the bottom (with a diffuser) can facilitate settling (see p. 68).

Separate inlets from two wells, and an overflow all the way at the top of a riveted, corrugated galvanized tank.

Caution: If both a well and a gravity flow supply a tank, the overflow must be high enough capacity to prevent gravity flow water at peak flow from raising the tank level to where it drains into the well. The check valve at the bottom of the line in the well provides some protection against backflow, but since these commonly leak it is not good design practice to rely on them for this function.

There should be shutoff valves, unions, bypasses, or some other way to shut off or divert the water supply to all inlets so you can service the tank.

Coupling weld detail. Note the use of a brass nipple (won't rust solid to tank threads) and galvanizing paint over the coupling and weld.

Well inlets, welded into the top of a galvanized tank. This inlet position enables the overflow to be at the top of the tank wall and still have no chance of cross-contamination (see p. 62).

Outlet

The outlet is the pipe through which you get water out of your tank. It should be as close to the bottom of the tank as possible without being so low that it sucks the settled muck off the bottom. This arrangement will maximize the useful storage capacity, helping you get your money's worth out of the tank.

You can create an emergency reserve (e.g., for fire) by installing a low outlet for the reserve, and a mid-level outlet for ordinary use.

There should be a shutoff valve on the outlet, to stop the flow of water in case of a massive leak, and so you can work on the system.

Storage in non-industrialized countries often doesn't have an outlet (or drain). This simplifies construction and all but eliminates the chance of catastrophic water loss. The water is taken out with a hand-held container, a bucket on a rope, or a pump.

This 6" outlet on a 50,000 gal *(190 m³)* bolted galvanized tank is a foot above the bottom. The spring water feeding the tank has a large amount of sediment, and this design allows for a generous layer of it to accumulate before the high flow through the 6" pipe will vacuum it up off the bottom and into the distribution system.

Service Access

There must be provision for getting inside the tank for inspection and cleaning. In the case of a tank that is so small that your arm can reach from top to bottom, it is sufficient to have an arm-sized opening. Otherwise, you need an opening of about 2' *(60 cm)*. For tall tanks, it is a convenience and safety measure to have a built-in ladder, at least on the inside. Rungs or stepping-stones should be spaced 1' apart vertically *(30 cm)*. An integral inside ladder eliminates the possibility of introducing contamination along with a portable ladder.

It is also convenient to have a pressure-tight service access door at ground level, which can be opened for construction and major service, to save the effort of climbing up the outside and down the inside.

If the tank (or a diversion) receives uncontrolled surface runoff, a manhole at the bottom can be a godsend for cleaning out truckloads of accumulated soil, rocks, and vegetation.

Placing the service access at the conical peak of this 13,000 gal (50 m^3) ferrocement tank allows several thousand gallons more water to fit before it overflows out the access.

Same tank inside. A ladder could have been incorporated in the central pillar, for safety and to avoid the hassle of feeding an extension ladder into the tank.

The access on this 50,000 gal (190 m^3) galvanized/poly bag tank is from the top only, as a perforation of this size of the liner would not be a good Idea. Attempting to fill the 8,000 gal (30 m^3) or so of space under the lightweight conical roof with water would be equally inadvisable.

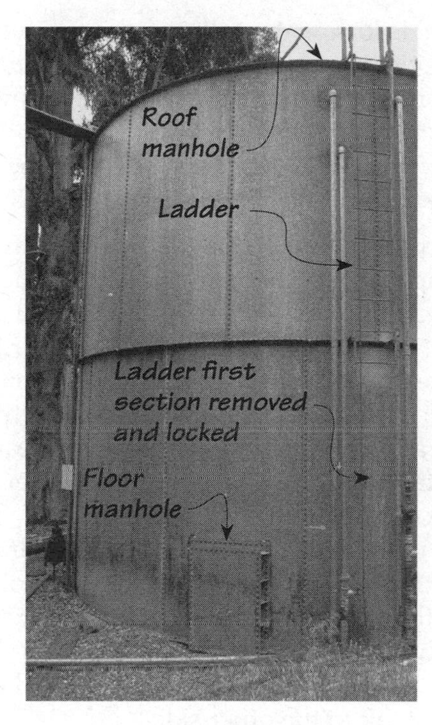

The access manhole at the bottom of a 20' (6 m) tall tank helps a lot for assembly, cleaning, and service. For access when the tank is full, there is a permanent ladder to a manhole at the top. A removable ladder section is locked elsewhere to discourage unauthorized access.

Ladder made of ½", epoxy-coated, grade-60 rebar integrated in a ferrocement water tank. This choice of rebar reduces the two main problems with rebar ladders: bent steps (grade 60 is half again as strong as standard grade-40 rebar) and rusting away, which can render the ladder unsafe as well as impact the structural integrity of the tank wall.

Drain

A drain retrofit using a 4" galvanized ninety. Since it was welded to the tank floor before assembly, there was no rubber gasket to worry about frying, and it could go right at the edge by the wall.

The drain is how you get the sludge, the last of the water, and the wash water from cleaning out of your tank. The practical consequences of a well-drained tank are:

- a tank that is cleaned much more frequently and perfectly by happier cleaners, leading to:
- cleaner water at the tap, especially when tank levels are low and the amount of muck entering the tank is high

The drain is the most neglected area of conventional tank design. We're going to provide many pages of information here to fill the void. What will happen if you ignore the drain issue? If your water is sediment-free, not much. If your water contains sediment, however, you'll be happy you took the trouble to learn about drains.

Tanks with No Drain

The classic tank design leaves the drain out entirely, so the *outlet doubles as the drain.* This brings such negative consequences as:

- When the level of settled sludge reaches the level of the outlet, it will suck sludge out into the pipes.
- When the water level is low, water falling from the inlet will stir up the sludge, occasionally yielding a concentrated sludge brew at the taps.
- Draining the tank completely for cleaning is an extremely tedious process using buckets, shovels, and finally sponges.
- There isn't an easy way to get water from washing the inside of the tank out.
- Bottom line: the tank doesn't get cleaned often or well.

What do tank manufacturers think? That their clients have totally crud-free water? Don't mind sucking sludge into their pipes? Have nothing better to do than remove a vast puddle of mucky water one sponge full at a time?

The 50,000 gal *(190 m^3)*, bolted, galvanized steel tank at right had no drain. The lowest outlet left more than ankle-deep sludge water on the bottom—several truckloads. It was an all-day affair with several muck-covered volunteers toting brooms, buckets, shovels, and sponges to get it halfway clean. Each successive rinse was an agonizing effort, and was indiscernibly cleaner than the last. Time and energy invariably ran out before the sludge did.

We partially remedied this sorry state by welding a 4" drain flush under the floor (photos and figure at right). This solved the problem of getting most of the sludge and water out. However, the absence of a slope to the floor still left the last little bit an exercise in frustration. You'd sweep a puddle toward the drain, and while the drain would intercept a narrow swath of the current, the rest of the water would just swoosh past and curve around the tank to the other side.

If these drainage problems had been anticipated in the original installation, we could have welded the drain right at the edge (before the rubber gasket was in place—see photo above), then installed the tank ever so slightly tilted, with the drain at the low spot. A tilt of ½% or 1% would be hardly noticeable, but would give tank-cleaners a huge edge; they could feed rinse water from the inlet (at the high side, of course) and sweep the increasingly clean water toward the drain at the low point.

This bit of tilt also could squeeze in another inch or so (a few hundred bucks' worth) of water level, by making more room to install the overflow at the high point.

All Stirred Up

A remote Coast Guard station in Cuba put my family up for a few nights. The only water was trucked in from many miles away, but there was no gas for the water truck. On the second day we ran out of water. The elevated water tank wasn't actually empty, it was just down to the level of the rusty muck in the bottom, a brew so thick and noxious that it wasn't possible to drink even when that was all there was.

Figure 16: Drain Retrofit For Steel Tank
Vertical section view

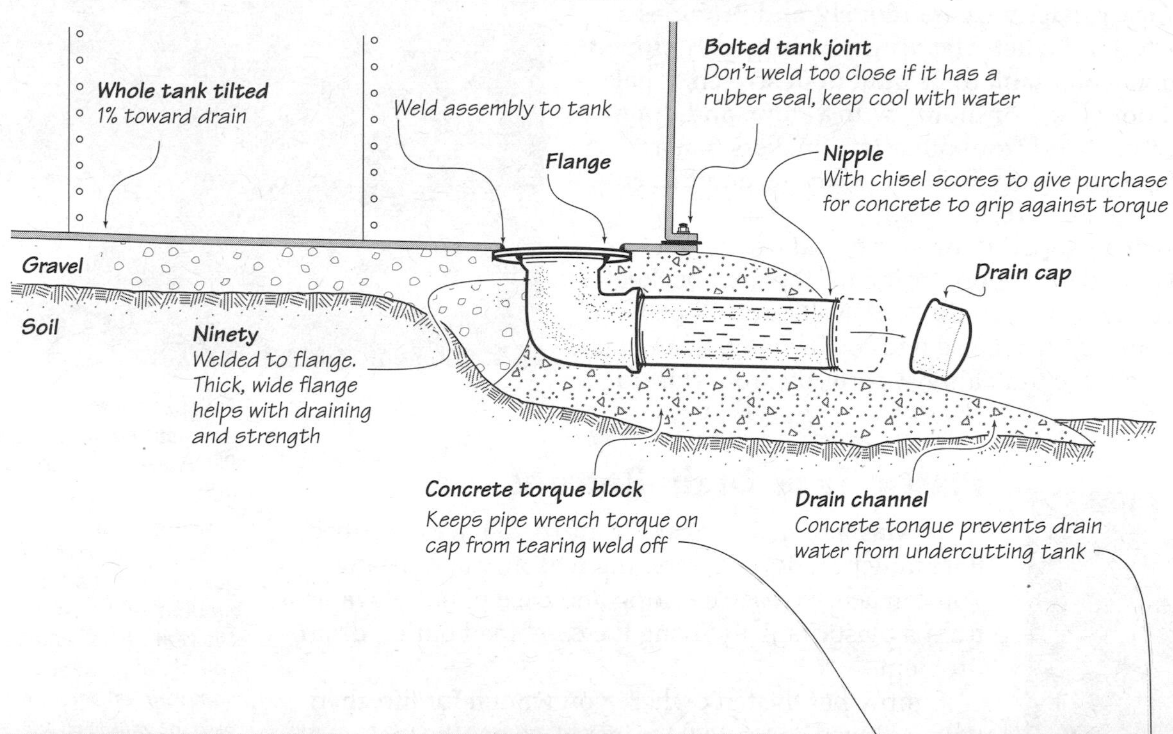

A 4" weld sweep ninety was welded to a flange and painted with galvanizing paint.

The whole assembly was then welded to the bottom of the old, assembled tank from the inside.

Welded drain from the outside, ready for plug and concrete work.

The completed drain was stoppered with a 4" plug (a cap would be better.) The concrete around the drain takes the pipe wrench torque from removing the plug, so the drain doesn't twist and tear out of the bottom of the tank.

Drain Location and Orientation

Some tanks come with a horizontal drainpipe at floor level. The capacity of such a drain is very low when the tank is empty. A big drain with a thin puddle of water on the floor has only a tiny part of its cross-section wet, so it can drain only a trickle of water.

What this means is, when you go to clean a tank, a wide, puddle will deepen until enough of the cross- section of the drain is wetted that it can accommodate the rinse flow. Without a sump, the drain flows slower and slower the lower the water level gets, and rinsing is still an exasperating operation.

Drain with Sump

A sump—a depression around the drain—increases the drain's low-water flow capacity tremendously, and provides a place for sludge to "catch" when cleaning a wide, gently sloped floor. When I design a water tank to be built from scratch, I make a distinctly sloped floor (1–2% or more), with a sump and drain at the lowest point. I've heard from other tank makers that a *flat* floor with a sump works very well for drainage, and this is clearly easier to make (Figure 18, at right).

To clean a tank with a sloped floor, sump, and drain, you just open the drain and turn the inlet on. Inside the tank, fill buckets from the inlet and splash water on the roof, walls, and sloped floor, then sweep them off, then rinse with more water. In an hour one person can clean a large tank almost perfectly, from start to finish.

This welded steel tank has settled onto fresh fill underneath, leaving it quite tilted. It has been this way for years and is working fine. This suggests that 1) a modest amount of intentional tilt won't hurt, and 2) you should always put the drain on the downhill side, especially if you've got fill under your tank, or it may not be at the low point for long.

Plastic Tank Drain Retrofit

In Masonry in and over Plastic (p. 47), we described the many benefits of combining masonry and plastic. You can achieve easy cleaning and cleaner outlet water from a plastic tank by fixing the combined outlet/drain problem.

Simply put the tank where you want it for life, then pour a sloped concrete floor inside, as per the picture at left and drawing below.

The water in this community drinking water tank in Mexico was forming a white, flocculent precipitate, apparently caused by sunlight coming through the black tank walls. The outlet, which doubled as the drain, was an inch from the bottom. It sucked unsightly crud out into people's bottles, yet did not allow for cleaning. Adding a sloped floor, drain sump, clean outlet, and plastering outside solved both problems.

FIGURE 17: RETROFIT OF PLASTIC TANK WITH A SLOPED FLOOR, SUMP, AND DEDICATED DRAIN. SECTION VIEW

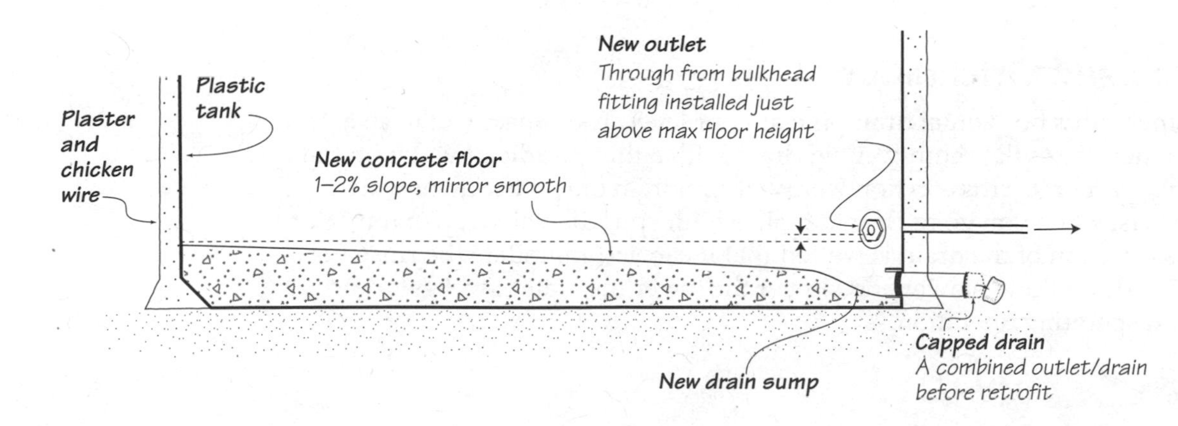

Bulkhead fitting: A fitting that can be inserted in a hole in the wall or floor of a tank, which has rubber washers on each side and a nut to tighten them for a leak-free connection

Figure 18: Drain Options

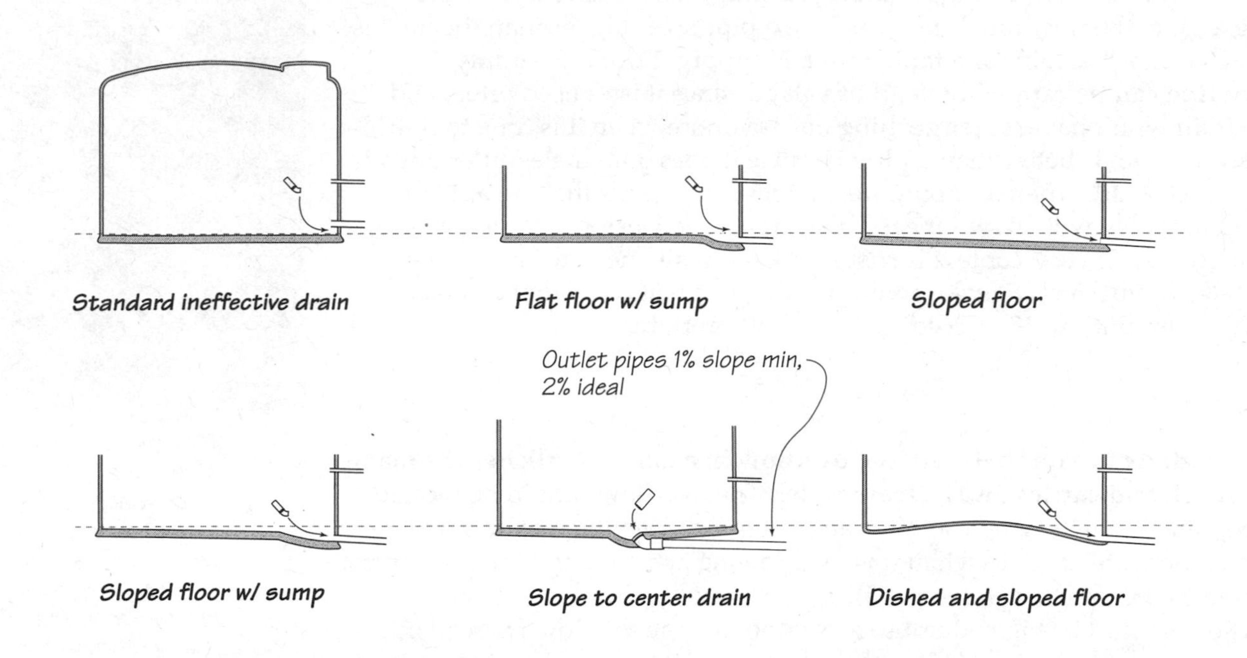

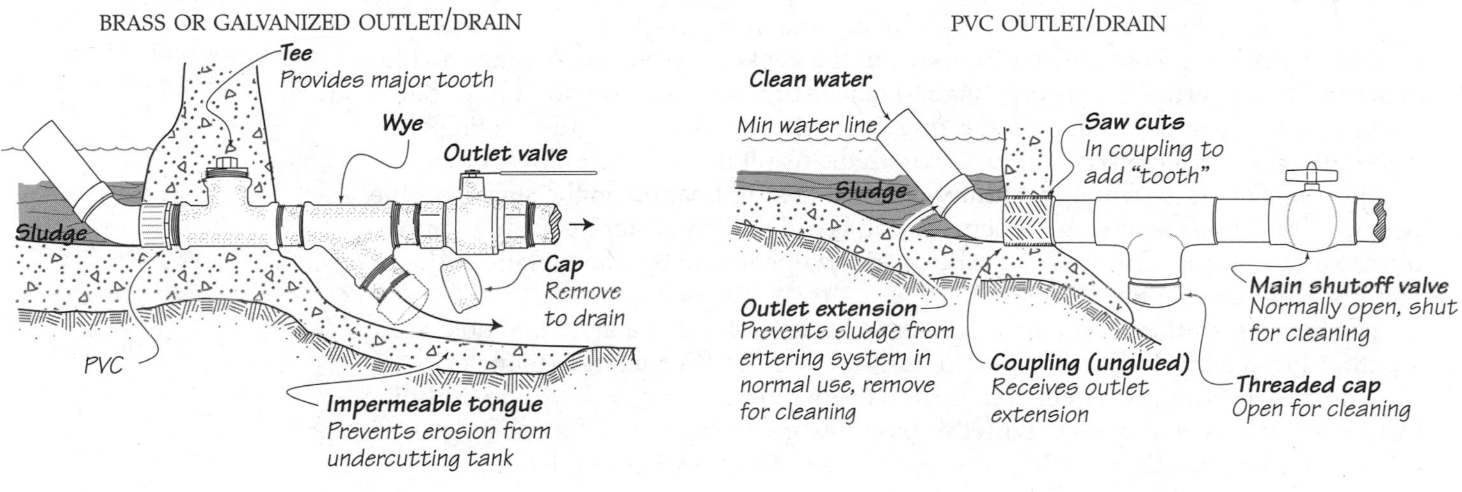

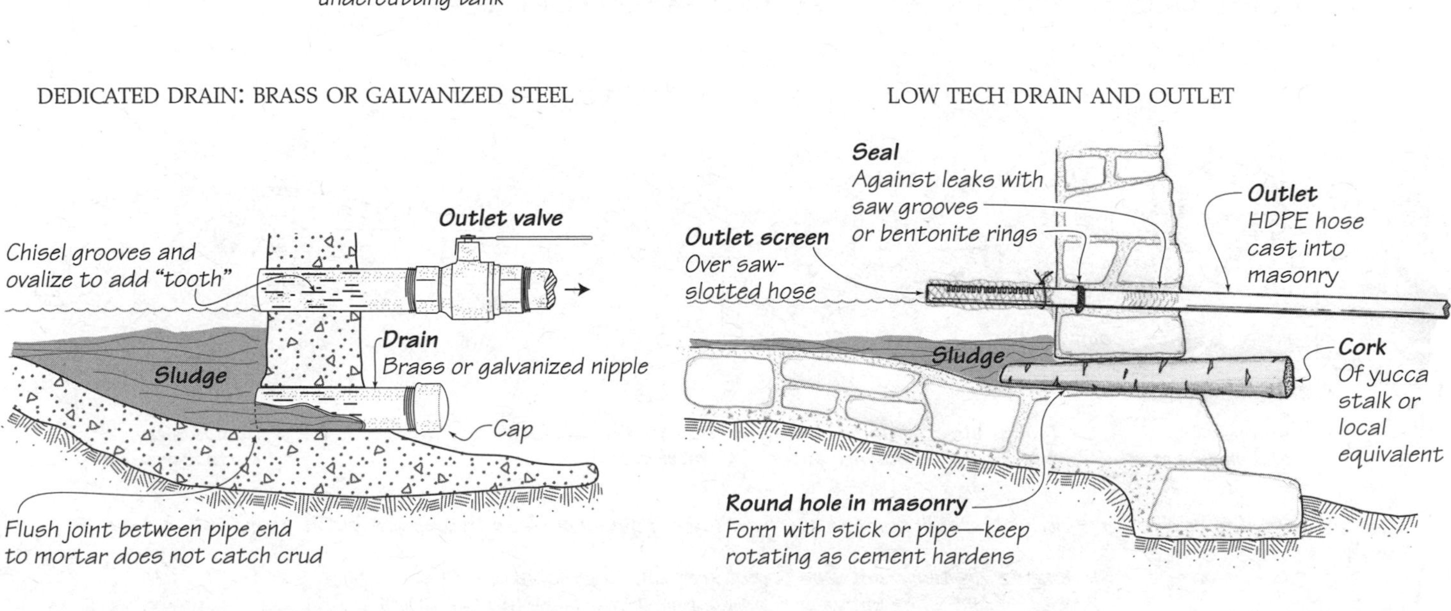

Drain Components

The drain should always be bigger than you think. With ankle-deep water pressurizing a pipe, the flow rate is very low. Two pipe sizes bigger than the inlet is about right. That is, a 4" drain for a tank with a 2" supply, 2" for a 1" supply.

The drain line can be capped instead of valved, since it isn't used often, and usually the drain won't have water gushing out any more when it is time to put the cap back on. A cap is better than a plug because it goes onto male outlet threads which won't catch crud, whereas a plug fits into female threads that are apt to fill with sand. A ball valve will allow for ease of opening and closing; a removable cap will save you money. If your context is rustic and you really need to save money, the flower stalks of yucca cacti make awesome plugs—giant corks that can take several feet of pressure (Figure 18, preceding page, bottom right).

This volunteer-built wilderness hot spring at Big Caliente has something every such pool should: a drain. Unscrew the cap, sweep out the pool, and refill it with clean water. The only refinements I'd suggest are a sump and a bigger drainpipe. This pipe is 1.5"–I'd have gone with 2" for faster, less clog-prone draining.

Overflow

Like the overflow in your bathtub, the overflow in a tank establishes the maximum water level, and carries away excess water. The overflow should be located as high as possible while leaving room for an air gap between the maximum water level (the level the water rises to when it is overflowing actively) and the inlet of any well connected to the tank (see Inlet, p. 55).

The overflow should be big enough to accommodate the full flow from all inlets. Usually this means it needs to be bigger than the inlets. With just a little pressure to push the water through, a 4" overflow might be needed to accommodate all the water gushing from a 2" high-pressure inlet.

The sizing of the overflow depends a lot on what happens if its capacity is exceeded. If water will just flow harmlessly out the vents or access hatch, that's no big deal. But if an overflow is the only way to relieve pressure and you don't have one or it clogs, you could end up pressurizing your whole tank as in Figure 32: Pressure Depends on Water Depth Alone (p. 91), with the result that the tank blows up.

The overflow is a great opportunity to be rid of stuff floating on the surface of the water. Why dump clean water when you can be rid of dirty water instead? To make the most of this opportunity, orient the overflow opening in the same plane as the water surface (see Figure 19, below).

The floating stuff rides on the top layer of water molecules, a layer that holds together like a tablecloth as it is pulled out the overflow. With a large-diameter, horizontal overflow, less water comes from deeper in the column, so you're dumping the dirtier surface water. Also, with this overflow geometry, the surface sheet gets tugged on more, pulling floating crud out of the farthest reaches of the tank.

Figure 19: Overflow and Inlet Height Specifications

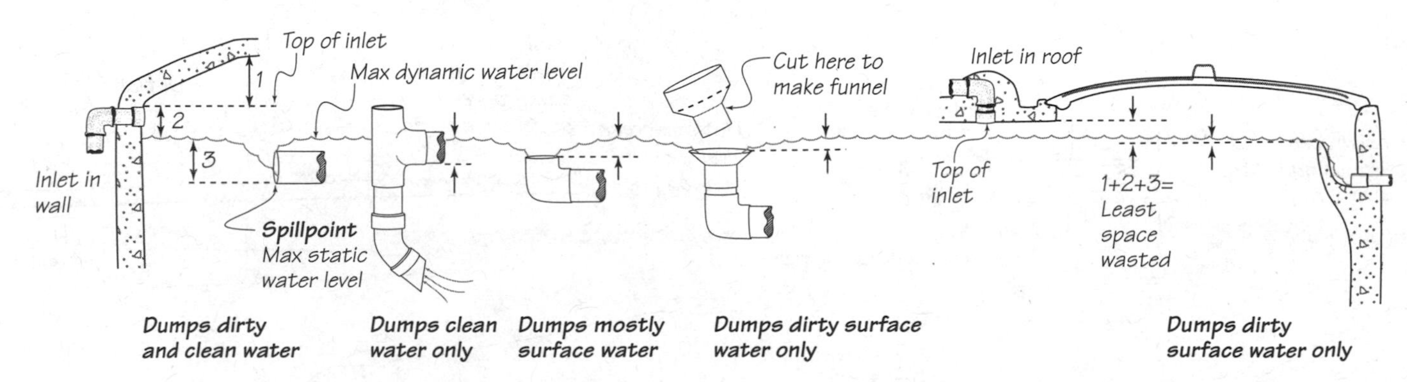

"Wasted space" is tank volume above the spill point of overflow. You've paid to enclose this space, but it stores air, not water. To reduce wasted space:

1. *Reduce distance between top of inlet and high point of ceiling*
2. *Reduce distance between max dynamic water level and top of Inlet*
3. *Reduce difference between overflow spill point and dynamic water level*

If the overflow opening is at a right angle to the water surface, a substantial fraction of the flow is drawn from lower in the water column, and you'll mostly be dumping clean water. If you install the overflow with its opening *entirely* under the surface, it will draw *only* from the mid-water column, leaving a surface covered with dust, leaves, and mosquito larvae.

If you use a large, horizontal overflow, you can place it slightly higher in the tank. An extra 2" of water depth in a tank 20' in diameter equates to 400 gal more storage.[m] This is worth at least $700, quite a bit more than the cost of the extra effort and materials.

The overflow should *not* have a shutoff valve on it.

12" (30 cm) overflow line on a 1.5 million gal, (5700 m^3) bolted steel tank.

A brass swing check valve, suitable for clog-resistant critter proofing of an overflow line. The swing check valve should always be installed in the overflow line in a horizontal position, so it swings shut; and some distance below the maximum water level, so the (slight) pressure needed to operate the valve doesn't cause the water level in the tank to back up higher.

Septic tank outlets have a tee on them, which specifically prevents floating solids from flowing out the outlet and into the leachfield. Pop off the tee, and the water level won't change, but instead of having a several-inch-thick layer of floating solids, there will be open water... and a clogged leachfield.

Exceptions—Now that you know all about overflows...you may not even need one. If the water source is above the tank, that itself protects it from contamination—an air gap below the inlet is not needed. If the water source is not expected to overflow extensively (for example, if you have a float valve or pump switch which normally shuts off the water supply when the tank is full, or if the supply is at a pressure which corresponds exactly to the maximum water height in the tank), the access port or air vent can double as the overflow, or the access port can "triple" as access, air vent, and overflow.

The side of this tank is rusted from water that overflowed the whole tank, prior to the installation of the overflow line (pipe at left).

[m]***Metric:*** *5 cm of depth in 6 m diameter tank adds 1.6* m^3 *of storage.*

Figure 20: Critical Overflow of a Rainwater Cistern Under an Office

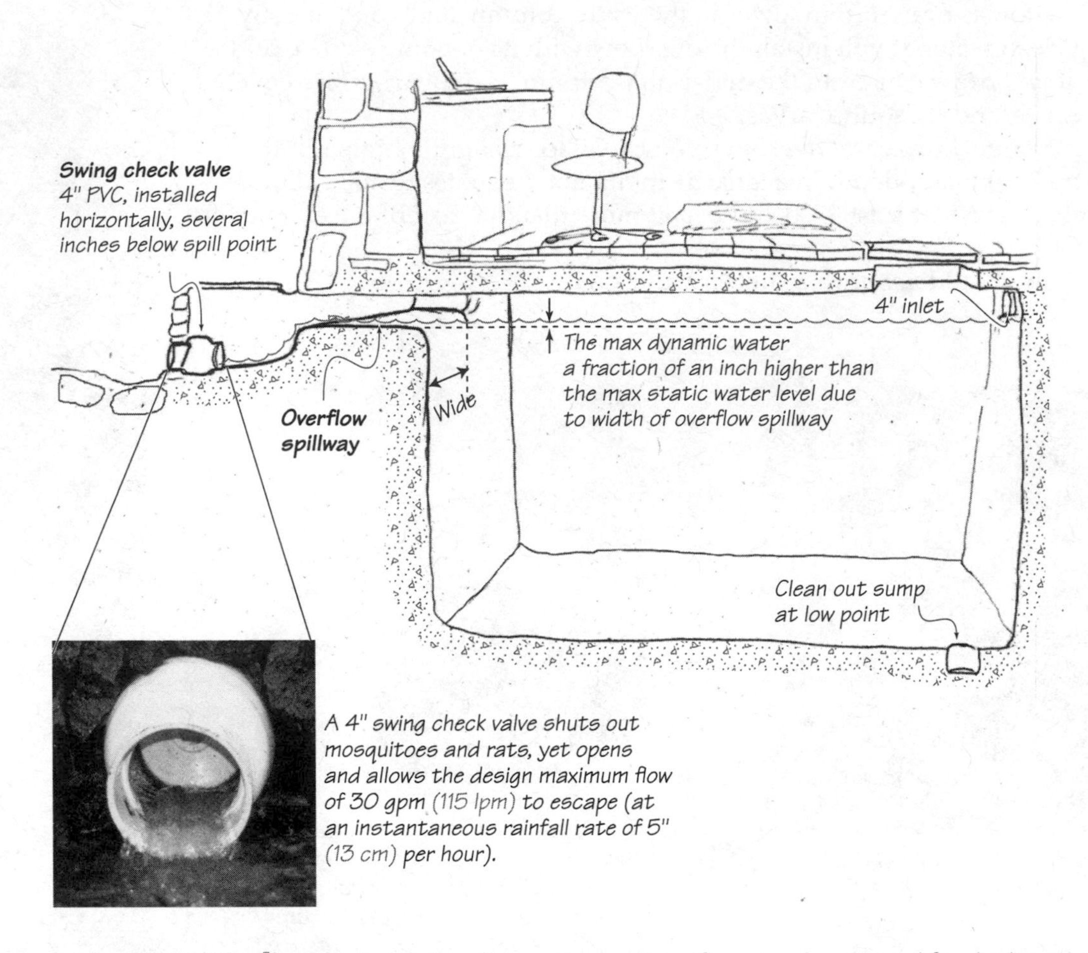

A 4" swing check valve shuts out mosquitoes and rats, yet opens and allows the design maximum flow of 30 gpm (115 lpm) to escape (at an instantaneous rainfall rate of 5" (13 cm) per hour).

A critical overflow on a 7,400 gal (28 m³) rainwater cistern harvesting system. The rainwater is used for drinking, so critter-proofing is critical. Moreover, if the overflow fails, rainwater would gush out of the tank access, over the hardwood floor, and through almost every room of the house. I sized the overflow very generously. The wide, horizontal spillway ensures that the level won't rise much under high overflow, while allowing the tank to hold a few cubic meters (several hundred gallons) more than if the spillway were a 4" pipe.

Uncontrolled Overflow

This 20 year old, welded galvanized steel tank has no overflow. Apparently the well, which supplies it with sulphurous water, has no tank level shutoff. Result: Water regularly overflowing from the access manhole has corroded through almost the entire circumference of the roof-to-tank joint. I probably could have ripped the whole roof off with my hands. The overflow caused further mischief on reaching the ground. An overflow-caused gully has grown from the nearby steep slope almost to the tank pad (at right). Just beyond the photo to the left, the gully is head deep. If the overflow continues unchecked, the whole tank will end up in the gully.

Critter-Proofing

ALL points of ingress to the system should be critter-proofed. Not only mosquitoes love water. Thirsty rats are particularly ingenious about getting into water tanks and dying there. There's nothing like pulling pieces of dead rat out of your water lines to get religion about the details of critter-proofing.

Block entrance points using mosquito net, welded wire mesh, closed valves, check valves, water seals, or a forceful outward flow of water. Water seals (like the traps on drain lines) will stop flying insects from getting in, but they won't stop rats. An overflow critter-proofed with mesh can clog, causing pressure to build in the tank, possibly exploding it. If you have an overflow that is too critical to restrict with wire mesh, a swing check valve is the answer (see Figure 20 on preceding page).

Quarter-inch welded wire mesh over gutters is first line of defense against rats crawling into this rainwater harvesting system. A water seal downstream stops mosquitoes.

Air Vent

If air can't get out of the tank, water can't get in. **There needs to be a way for air to get in and out of the tank, one which is screened against insects and rodents getting in, and doesn't admit much light** (see Sunscreen and Shade, below).

The vent geometry should preclude runoff from the tank roof from entering. You should generally only have as much venting as is needed to make way for the incoming water, which isn't much at all. If you have a huge airflow, you may lose a significant amount of water to evaporation.

Sunscreen and Shade

Sun plus water equals the base of a food chain. Almost all water sources have enough trace concentrations of nutrients that if you put them in the sun long enough, some algae will grow. Then something will want to eat the algae, and so on...

Metal and masonry tanks completely block sunlight. Plastic tanks should be black and/or indoors, buried, or covered, to prevent light from getting in the water. (See Masonry in and over Plastic, p. 47.)

There is every advantage to keeping water cooler in the summer. You can put the tank in the shade and paint the outside of it white, silver, or an unobtrusive, light version of the local rock or vegetation color scheme. This will lower water temperatures and thermal expansion/contraction stresses in the above water portions of the tank. *(Note: A black plastic tank painted white will be both cool and dark inside. An unpigmented, cloudy white plastic tank will quickly degrade and algae will grow vigorously inside.)*

Critter-proof service access, air vent, and backup overflow all in one. Note that vents around access are screened with ¼" hardware cloth over mosquito net.

Chapter 5: Optional Water Tank Features

In this chapter we go over various optional gadgets which can enhance the function of your system.

Be aware, too, that there are a host of other possible options which we're not going to get into (remote flow meters, automated injectors for treatment chemicals, online monitoring meters for turbidity, chlorine level, etc.).

Inlet Meter, Filter, Gauges

If you want to optimize the use of a resource, measure it. Our water system has two wells and a spring, each with a meter on the inlet, just before the water tank. Each house that is connected to the water system has a meter. With regular readings, these meters provide information valuable for the management of the system, such as the amount of water capable of being produced by our wells and spring in wet and dry seasons, consumption in wet and dry seasons, etc. My next house is going to have a water meter set into the tile above the kitchen sink to keep us continuously apprised of our water consumption.

An inline filter protects a meter or a float valve on the inlet. Place it just before the meter or valve, to keep their mechanisms from clogging with chunks of rust from the supply line.

Place an inline inlet filter just below the "breakaway" section of the line from a creek direct system—that is, just below the point where the line is threatened with floodwaters. If the filter in the creek gets knocked off, this inline filter will keep the whole line from filling with gravel.

Inlet hardware. Pressure gauge before shutoff shows static pressure. Pressure gauge after shutoff shows dynamic pressure across filter, which shows when it needs service.

Pressure gauges can tell you the level of water in the tank remotely, the operating or static pressures in supply lines, etc.

Float valves automatically shut off the water when the tank fills.

Inlet Float Valve

A float valve on the inlet is useful for shutting off the flow from a gravity flow water supply when the tank is full. This not only avoids the pointless removal of water from nature, it will improve the quality of water in your system. One water system I worked on didn't have a float valve for years. Every rainfall, the springs would flood and the inlet would gush muddy water full-blast into the tank day and night. Most of the muck would settle in the tank, while clean water poured out the overflow.

Inlet Combined with Outlet

Under some circumstances it is desirable to combine inlet and outlet functions in the same line, through which the water flows both into and out of the tank. This is most advantageous with long lines. Instead of having two lines, you have just one, which splits at the tank into inlet and outlet forks (see Figure 21).

An example of this application would be a gravity pressure tank—a tank that water is pumped up to, and from which it runs back down into the system. This is frequently the function of water towers and they are often plumbed this way.

If the gravity tank is connected directly to a well (as in Figure 21), there is some trade-off in water security. If the well pump check valve leaks, the tank can drain back through the leaking valve into the well. If the water in the tank happens to become contaminated while the check valve is leaking, the well can get contaminated also. Discharging the well into a storage tank at wellhead level, and pumping to a higher storage tank from there can circumvent this issue, which can be legal as well as practical.

FIGURE 21: INLET COMBINED WITH OUTLET

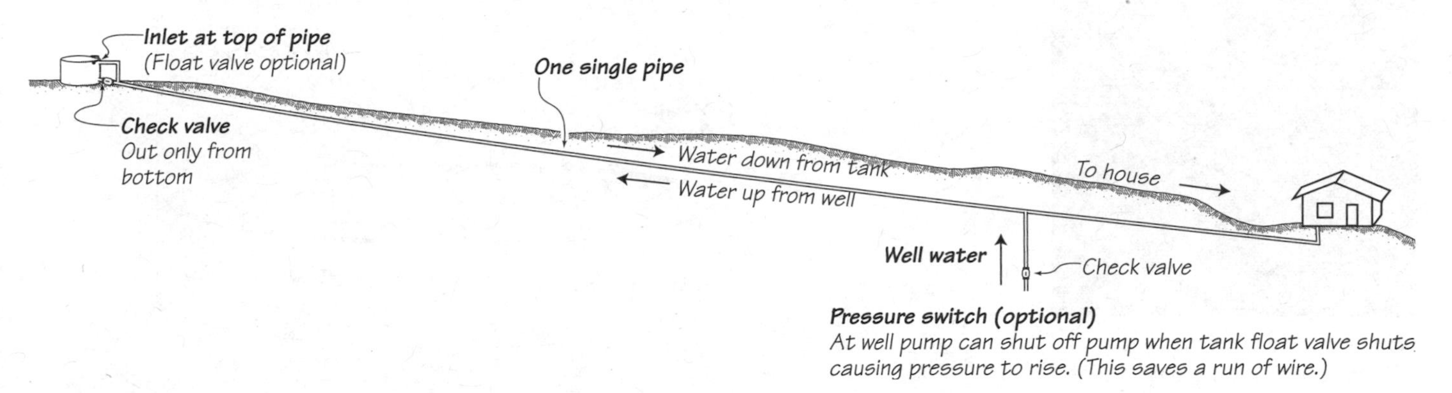

Inlet Aerator

Aeration is the process of breaking water into fine droplets mixed with air. This creates a large surface area-to-volume ratio. Oxygen can readily dissolve into the water, and gasses dissolved in the water can readily escape. Certain kinds of water quality problems can be improved by aeration; for example, manganese and iron oxidize to a less objectionable form.[28] Noxious gasses in the water, such as hydrogen sulfide and chlorine, can also be driven off more rapidly by aeration.

If there is enough pressure, the inlet can be aerated by capping it with a sprinkler or mister that thoroughly mixes the water with the air in the top headspace of the tank. If there is little pressure, the water from the inlet can be broken into droplets by passage over a tower of tiered screens or slats.

For aeration to be optimally effective, there needs to be enough ventilation to move oxygen in and undesirable gasses out of the tank.

Inlet Diffuser to Improve Settling

Turbidity (suspended solids) generally settles out with storage in tanks.

One community I worked with was faced with a tightening regulatory noose which was requiring them to meet a very stringent turbidity standard for water from their springs. (Less than 1 NTU, which equates to about 10' visibility).

We replumbed their two 50,000 gal *(190 m^3)* storage tanks to reduce the spring water's turbidity as much as possible. (See Figure 22, Inlet Diffuser, next page.) Instead of letting the water level fluctuate in both tanks, I dedicated one as the spring water settling and treatment tank. Before, the spring water was introduced at the top, with the hope that the turbidity would settle out before reaching the outlet (also at the top). Instead, we ran a new line from the inlet float valve down to a ring of pipe knee height from the bottom, with finger-size holes drilled in it every few feet.

The water (and suspended solids) exit in a diffuse ring around the entire bottom of the tank, then flow ever so slowly up the entire water column of the tank. Arranged thusly, the suspended solids start almost at the bottom, where we want them to settle, and would have to float up nearly 20' against the force of gravity to get to the outlet, something they are unlikely to do.

FIGURE 22: TANK OPTIONAL FEATURES: INLET DIFFUSER, OUTLET FILTER, VARIABLE HEIGHT OUTLET

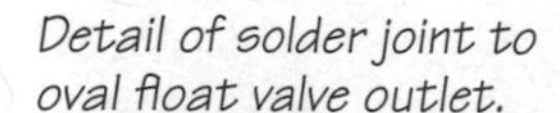
Detail of solder joint to oval float valve outlet.

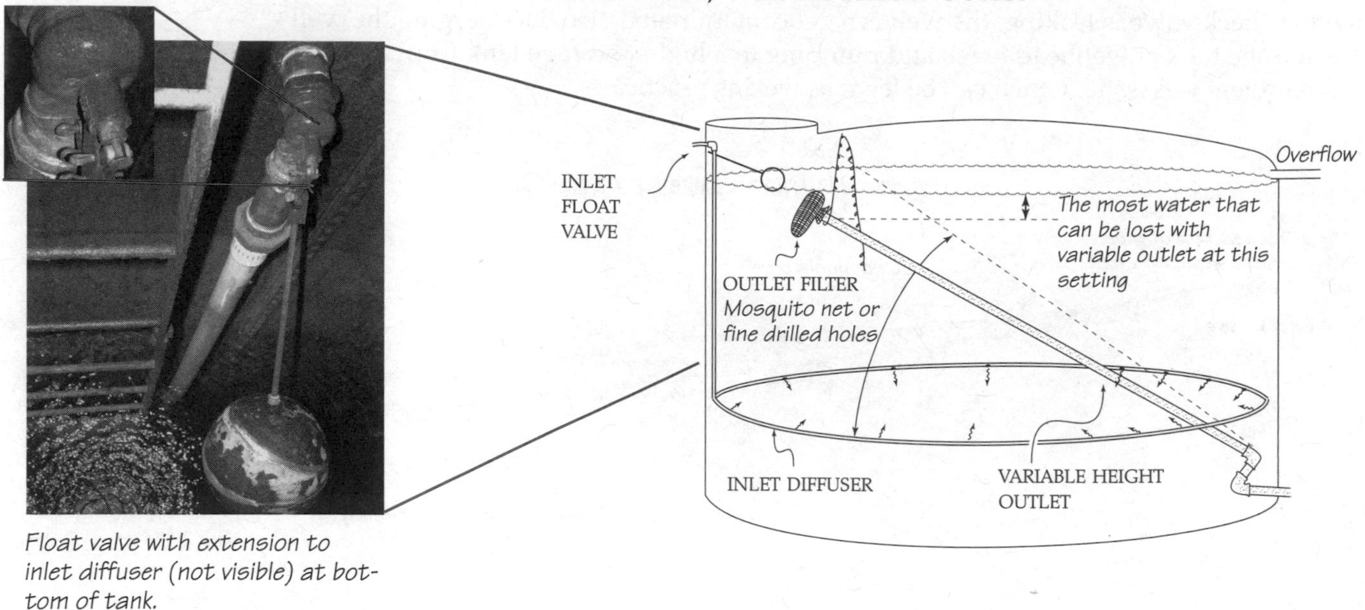

Float valve with extension to inlet diffuser (not visible) at bottom of tank.

The outlet of the settling and treatment tank is almost at the same level as the overflow, so the tank is always full. Thus, there is never a low-level situation where the inlet water is plunging through the air to violently stir the muck on the bottom. In terms of storage, the system now provides an emergency reserve which can be accessed by opening a valve, and is secure against accidental loss from a catastrophic leak or unnoticed supply interruption.

Outlet Screen or Filter

If there is reason to believe that something might come out of the tank which could clog the lines or valves downstream, it is cheap insurance to have a screen over the outlet. A clog will be easier to deal with there than elsewhere in the system. If getting into the tank to service it is too much hassle, you can use an inline filter just outside the tank.

Variable Height Outlet

If your water supply is so tight that accidentally draining your tank would be a disaster, consider installing a variable height outlet. This is a flexible or hinged extension to the outlet on the inside of the tank, which can be manually raised or lowered to track just below the water level in the tank, so there isn't much water above the outlet. If there is a leak or some other problem, the only water that is vulnerable to loss is the volume between the surface and the level of the outlet. (See Figure 22, Variable Height Outlet.)

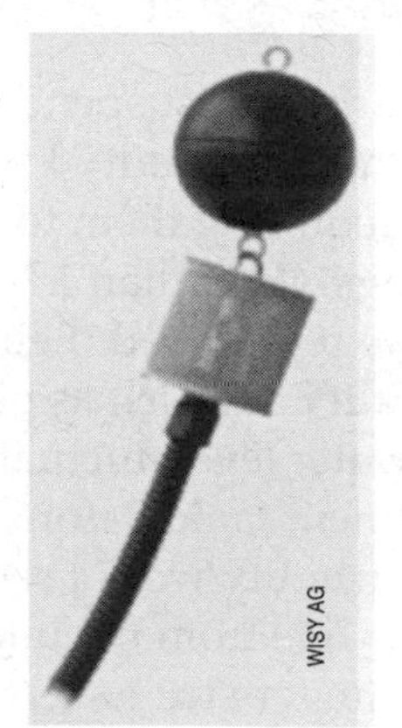

Flexible outlet float.

Outlet Float

The water in a tank is generally cleanest about 6" *(15 cm)* from the top of the water. You can take water from this level by extending the outlet on the inside of the tank with a flexible line and float.

Water Hammer Air Cushion

If you have a long water line with a high rate of flow and someone slams a ball valve shut at the tank inlet, the moving water column is like a battering ram impacting a closed door. The resulting pressure spike can be many multiples of the static pressure and can easily blow the pipe up. To avoid this scenario, you can:

- use gate valves, which are not possible to shut suddenly
- use high-pressure pipe
- put big warning signs on the valves in several languages
- install air cushions

An air cushion is simply a part of the plumbing that traps enough air so that when the water hammer pressure spike hits it, it cushions the blow. You can buy bladder-filled ones or make your own (see Figure 23, above). Put them in a place in the system where they will be self-draining when the system is drained. Otherwise, they will eventually fill with water, reducing their effectiveness.

FIGURE 23: WATER HAMMER AIR CUSHION

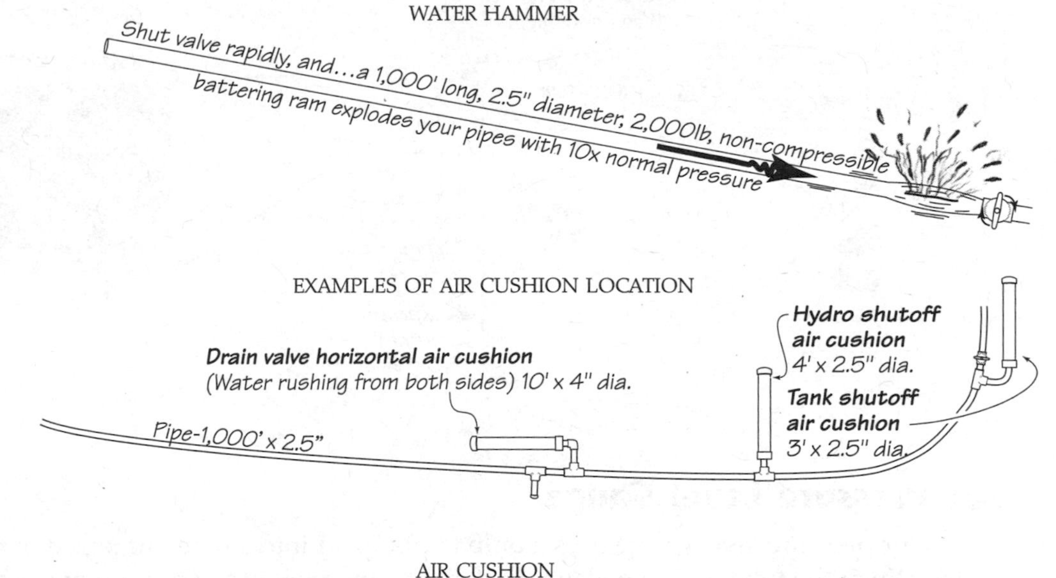

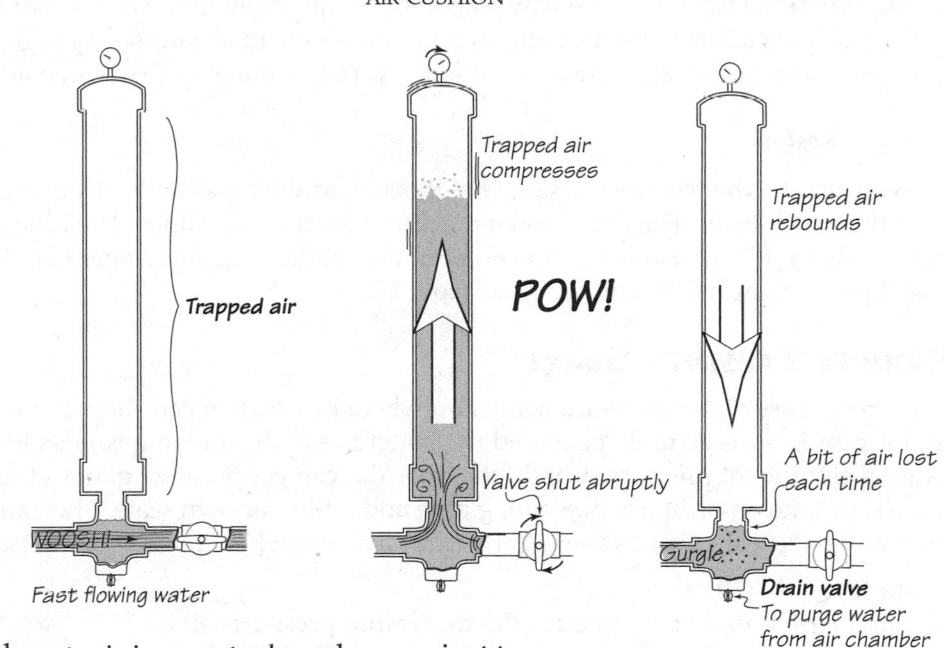

Level Indicators

It is reassuring to know how much water is in your tank, and convenient to be able to determine this without having to climb to the top and look in.

There are several gadgets that can help you accomplish this:

Float

A float in the tank raises and lowers a weighted marker on the outside of the tank. These common devices are simple and reliable, so long as the weight and float are heavy enough to keep the rope or cable that connects them from sticking where it passes through the tank (see Figure 24, Level Indicator, next page).

The marker can be positioned so that it can be seen from far away. Note that when the marker is at the bottom of the tank, the water level is at the top—this takes a bit of getting used to.

Figure 24: Tank Optional Features: Level Indicator, Ozonator

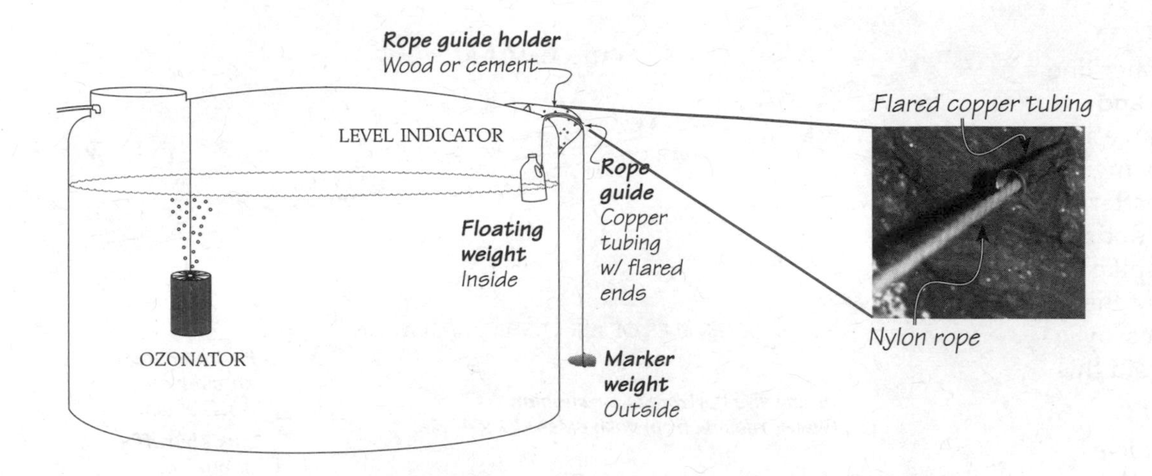

Bought float-type level indicator.

Air Pressure Level Gauge

An air pressure level gauge uses a bulb to pump air into a tube immersed in the water. When the tube is all the way full of air, the pressure of the column of water trying to push back into it can be read as water depth on an air pressure gauge. With the line and gauge full of air (instead of water), there is nothing to freeze in cold climates.

Air pressure gauge to show tank water level.

Clear Tube

A clear tube that covers the same height span as the tank can be used to directly read the water level. The tube can be marked with reference lines. The tube must be either filled with chlorinated water, kept in the shade, or drained when not in use, or it will grow algae and become difficult to read.

Remote Pressure Gauge

An ordinary water pressure gauge anywhere in a system can show the water level in the supply storage tank, provided the water use at the moment isn't so high as to cause a significant pressure drop in the line. You can get pressure gauges that read in inches, or take an ordinary pressure gauge and add your own scale. The gauge will do a more accurate job of showing the exact water level in tank if it has these features:

A big face—4" *(10 cm) is good.*

A range which just encompasses the maximum pressure—*If the pressure with a full tank is 54 psi, a 60 psi gauge will give a better reading than a 100 psi gauge.*

High precision—*±1% or better.*

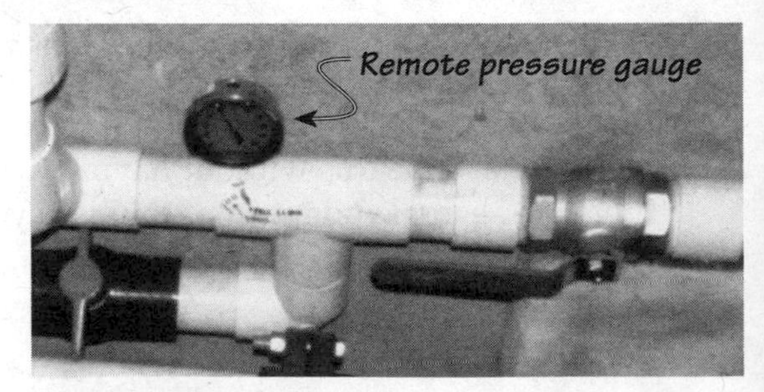

Precision pressure gauge shows water level at a tank 1,000' (300 m) away and 60' (20 m) higher in elevation to an accuracy of 1" (2.5 cm). This enables flow to be determined indirectly. From zero flow to 200 gpm (1 m^3 every 3 min) the water level rises in the overflow channel (and thus the whole pool) by about 4" (10 cm). The same gauge also serves as the dynamic pressure gauge used to check the flow adjustment to the hydroelectric wheel connected to the system.

Electronic Level Indicator

A bought electronic level indicator can give a remote reading anywhere you can run wires or a radio signal.

Calculate Gallons Per Inch

All the gadgets above will give you readouts of water depth. To translate this into water volume, you'll need to measure or calculate the gallons per inch of water depth *(or m3 per cm)*. You can use our *Water Tank Calculator*[6] to help calculate the volume. If your tank is a shape that is too strange to calculate, empty or fill it through a water meter, and make a mark at the water level that corresponds to each increment of volume. Knowing how much water use (or production) is represented by a given change in water level will sharpen your monitoring of the system.

Ozonators

Ozone is a great alternative to chlorine for disinfecting water. (See Hazardous Disinfection Byproducts, p. 10, for reasons not to use chlorine.) The way ozone treatment typically works is by generating ozone with UV lamps or corona discharge equipment, then pumping it through a diffuser or diffusers in the tank, or injecting the ozone directly into a tank feed line.

The water tanks are an integral part of such a treatment system. A tank full of water saturated with dissolved ozone can handle spikes in the amount of incoming debris and/or pathogens, whereas the low, steady output of the ozonator by itself could easily be overwhelmed. (See photo at right and Figure 24, Ozonator.)

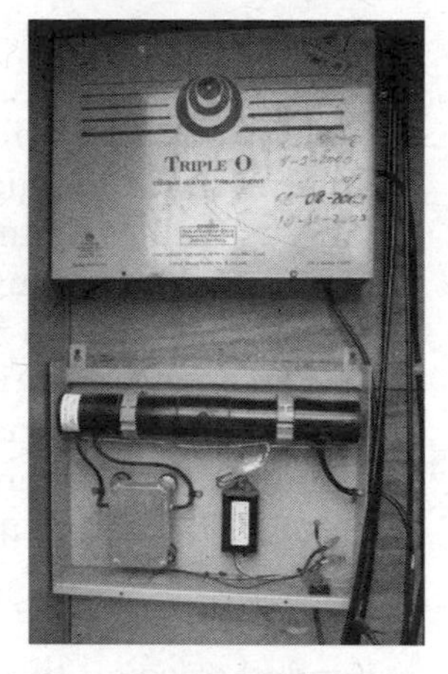

Ozonators provide sterilization with far fewer environmental and health side effects than chlorine. The ozonators shown treat water for about 30 homes.[30]

Drain Extension or Baffle

An unglued extension from the drain to a level above the sludge in the bottom of the tank will enable the drain to be used as both drain and outlet. With the extension pipe slipped in place, the water leaves the tank above muck level—a clean outlet. Pull the extension out, and the muck will suck out of the drain.

In the case of an outlet on the side of the tank, a baffle could deflect its suction upward, so it doesn't suck crud off the bottom of the tank. (See Figure 18: Drain Options.)

Outlet and Overflow Curves

Sharp angles and transitions at an outlet or overflow create turbulence that limits the peak flow. If you form gently curved transitions that smoothly funnel the accelerating water flow into a narrower space, you can get a higher peak flow. The ideal shape is like a graph of the water speed; wider where the flow is just starting, narrowing as it speeds up, and then constant as it's made it up to speed and into the pipe. There is a photo of a very successful curved overflow for a runoff diversion dam on p. 25 and drawings of flared overflows on p. 62 and 64.

FIGURE 25: OUTLET CURVES
TO LOWER TURBULENCE AND INCREASE CAPACITY

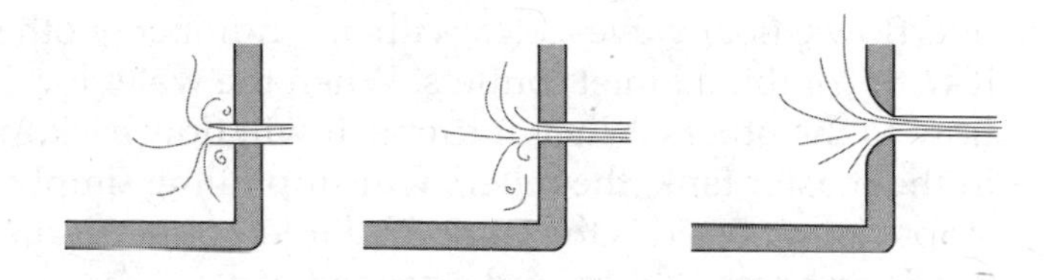

The smoother the water path, the less turbulence and the more water can flow out a given size outlet or overflow.

Pump Controls, Alarms, and Switches

The simplest pump control is to manually turn your pump on, and when you hear or see water pouring out the overflow, shut it off.

- **You can install a float switch which will turn on your well pump automatically when the water reaches a specified level**—Turn it off again when the tank is full. The switch can be hard-wired to the well pump, or connected via a radio transmitter.
- **An alternative pump control system uses a special float valve and a pressure switch**—The special float valve shuts off the inlet all at once when the tank is full. Pumping against the closed valve, the pressure on the line will spike upward, and a pressure switch at the pump turns it off.[29] Encase the float switch in a Ziploc bag to keep bugs out of it—a common cause of failure.
- **A low-level alarm switch**—Turns on a light or an audible alarm when the water drops below the specified level.

Sand Filter

In a slow sand filter, the water passes through a layer of sand from top to bottom. Treatment is biological and mechanical. Beneficial bacteria form a thick film on the surface, as well as a film over the interior sand particles. Suspended particles catch in this layer, and pathogens are eaten by the beneficial microscopic organisms. A properly made and managed sand filter has very high pathogen removal rates.

Slow sand filtration is a simple, inexpensive technology for treatment of water that may have pathogens.[31] It is especially appropriate for rustic homes, villages, and small communities that are required to use filtration to comply with new regulations. (See Figure 30: Small Sand Filter, p. 86.)

Emergency storage and sand filtration can be combined in one big tank. If the peak demand for filtered water is greater than the sand filter yield, then a separate tank is needed to cover these peaks (this is usually the case). Flow rates in slow sand filters are slow: ⅕–⅓ gal per hour per ft^2 of filter surface area *(7–11 L per hour per m^2)*.

Multiple Tank Management

There are plenty of reasons to have multiple tanks—different kinds of water, specialized treatment or settling functions in particular tanks, emergency water set-asides, or simply that you bought one tank and then another. Each of these reasons, in each context, will have its own optimal plumbing layout. Often, the optimal number of tanks is two, of equal size. This can facilitate maintenance and give flexibility of configurations.

When using multiple tanks, the key to simplified management is to **install them so that the maximum water level in all tanks is the same.** This reduces the redundancy in inlets, outlets, and controls tremendously. (It may not be a problem to have the floors of different tanks at different levels, provided the master tank has the lowest floor—see Figure 26, opposite.)

At the extreme, one tank alone can have the master inlet and outlet, level indicators, overflows, float valves, etc., with any number of other tanks plumbed to the bottom of it with combined inlet/outlets. When the water level rises in the master tank, water will flow to the others. When it drops, it will flow back. When the water reaches the overflow in the master tank, the others will stop filling simply because the level in the master tank stops rising. Besides the combined inlet/outlet at the bottom, each "slave" tank only needs an access, drain, and screened vent.

Connecting to a second tank with a maximum water level a foot lower resulted in the loss of 5,400 gal of useful capacity—thousands of dollars worth—in the first tank (at right), as the overflow had to be lowered to avoid having to duplicate all level-regulation hardware.

A tank farm for a rainwater harvesting system on San Juan Island, Washington State. The tanks are connected in series at the bottom, so they all rise and fall at the same level, and the water does not stagnate.

Figure 26: Plumbing Options for Multiple Tanks

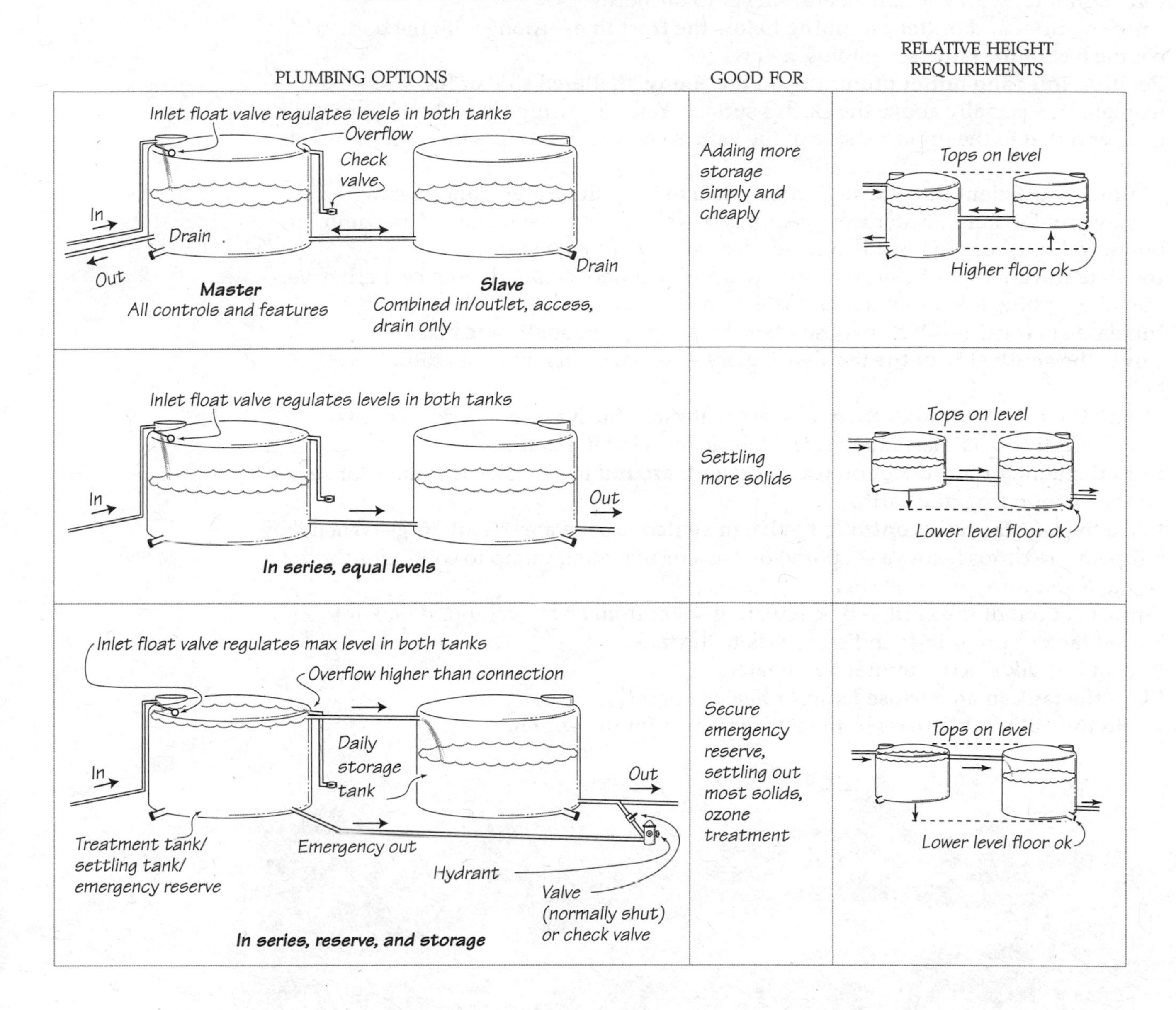

Freeze Protection

Water has high specific heat. It absorbs great quantities of heat and stores it for a long time. It reacts slowly to temperature variations. It takes a lot of time and energy to heat water, and it must lose a lot of heat energy before it freezes. In climates with wide daily temperature swings, a tank of water can keep itself from freezing by virtue of its thermal mass alone.

Besides the obvious benefits of keeping water from freezing solid, there can be an advantage to keeping stored water warm. If the water is to be used indoors, warmer incoming water reduces the heating load for both the building and the hot water heater.

Following is an inventory of methods to provide freeze protection for stored water Some can be mixed and matched; others are mutually exclusive. As always, the passive methods (listed first) are cheaper and more reliable. For really extreme cold, you will want to research other techniques.[32]

- **Place the tank in a warm microclimate**—Exposed to the southwest (to heat it up before nightfall), with windbreak and shelter to the north.
- **Bury the inlet and outlet plumbing below the frost line**—Along with the bottom couple feet of the tank (see photos, below).
- **Position inlet and outlet plumbing on the sunny, sheltered side of the tank**—And insulate it, especially above the earth's surface. You can ensure circulation by extending inlet or outlet to the opposite side of the tank, so new water moves through the whole tank.
- **Maintain sufficient flow through the system to keep the water from freezing**—The turnover of incoming water from a spring or well may not seem warm for swimming, but in freezing conditions it is warmer than the water in the tank.
- **Insulate the tank**—With strawbales, sawdust, or, if underground, pumice, perlite, vermiculite, or blue foam foundation insulation.
- **Increase solar gain**—Of an exposed tank by painting the south side black.
- **Cover the south side of the tank with glass**—To make it a low-temperature solar heater.
- **Shield the tank with high thermal mass material**—Such as stone, brick, or adobe (most effective in climates with large daily swings in temperature).
- **Bury the tank, or build a berm (earth mound) around it**—To take advantage of heat and insulation from the earth.
- **Use a thermostatically controlled valve or switch to set water circulating**—When the temperature drops below a set point, or use a recirculating pump to continuously circulate water through the system.
- **Draw heat out of the earth**—By circulating water from an aboveground tank into a buried tank or pipe, and sending it back to the tank.
- **Use an in-tank electric immersion heater.**
- **Place the tank in an enclosed and/or heated space.**
- **Drain the vulnerable parts or the entire system for the winter.**

The floors of these ordinary HDPE rainwater harvesting tanks in Santa Fe, New Mexico, are buried 3' (1 m) underground. This protects the delicate outlet plumbing from freezing, but it is not so deep that the tanks collapse from soil pressure.
The little box at right houses the pressure pump, which is prevented from freezing either by draining it for the winter or by heating with thermostatically controlled incandescent light bulbs.

Chapter 6: Emergency Storage

Storage for emergency supply can come from:

- secure reserve in your regular tank
- reservoirs of water in your house plumbing
- alternative sources in your surroundings
- longterm storage in sealed containers

After considering how much water you need, we'll run through these options, then look at how to protect stored water, and finally, water storage considerations for fighting fires.

How Much Emergency Water Do You Need?

For drinking and cooking, 2 L per day per person is a reasonable figure. With fanatical conservation, all other uses except clothes washing can be met with 5 gal a day *(20 L)* of water, or even less. Clothes washing is resistant to conservation—it will probably take as much water as everything else combined. Since any old water will do for clothes washing, you're better off planning on going to the river for laundry than storing the water. (If you find yourself suddenly constrained to this level of water consumption by disaster, you can take solace in the fact that this is about how much water humans have lived on in most areas for all of history, including most people today.)

In Kenya I found ten-person families using five gallons of water per day—total!
—Ianto Evans

How long a time period to store water for is another question. How long do you expect the water could be off? For most contexts, a week to a month's worth of water is a reasonable amount. In some places with a distinct dry season, it may make sense to have enough stored water to make it through the dry season, even if it is eight months long.

Emergency Storage You Already Have

The regular storage in your water system can serve well as emergency storage, especially if you pay attention to the water security suggestions in the next section and elsewhere. (Plumb your tank with an emergency reserve, Figures 15, 29, make your tank structurally sound, Appendix B, etc.) The advantage of using your regular water storage for emergencies is that it is maintained automatically as a consequence of the regular maintenance of your system.

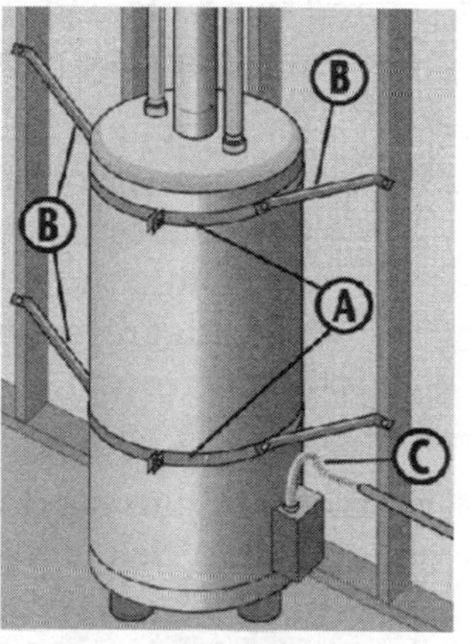

Earthquake strap for water heater: Wrap a 1-½" wide, 16-gauge-thick metal strap (A) around the top of the water heater and bolt the ends together. Do the same about ⅓ of the way up the side of the water heater. Take four lengths of EMT electrical conduit, each no longer than 30". Flatten the ends. Bolt one end to the metal strap as shown (B). Screw the other end to a 2" × 4" stud in the wall using a 5⁄16" × 3" lag screw. Be sure a flexible pipe (C) is used to connect the gas supply to the heater.

Chances are that a bit of resourcefulness will turn up a considerable amount of alternative water supplies also, for example:

- **Water in your plumbing**—Open the highest tap and collect water from the lowest.
- **Ice cubes in the freezer.**
- **Water in the toilet tank**—(Not the bowl!) is worth saving and using. You can unhook the flush valve to keep someone from flushing away this suddenly valuable resource.
- **The hot water tank**—Is a substantial reserve of usable water. It can be accessed through the drain valve at the bottom. The water may come out with calcium deposits in it, which can be allowed to settle out before use. Hot water tanks should be secured with metal straps so they don't tip over in an earthquake.
- **Rooftop solar heaters**—Can yield emergency water, too, especially if they are of the storage type. If the plumbing in the house is trashed, you could shut off the service shutoff valves to keep water from leaking out of the solar heaters. Usually there will be service shutoff valves on both the incoming and outgoing lines.
- **Hot tubs, pools, and decorative ponds**—Contain a large amount of water suitable for washing, bathing, and firefighting.
- **Nearby surface water**—If you have a water source nearby such as a natural watercourse or reservoir, you could supplement your emergency stored water with water carried from that source by hand, wheelbarrow, bike cart, car, or truck. You are fortu-

nate if you are surrounded with water that is naturally drinkable. But, as billions of people worldwide are aware, even quite funky water serves fine for most uses (see Table 1: Different Water Qualities for Different Uses, p. 7).

- **Rainwater**—Collected directly from the sky into bowls is OK to drink directly virtually everywhere. Rainwater collected from roofs can have pathogens from rooftop critters such as rats, raccoons, and monkeys. Rainwater is exceptionally well suited to use for washing, dishwashing, and bathing.
- **Nearby groundwater**—Indigenous people, forced to find water in your suburb, might tap into an abundant, fairly clean supply accessible via a shallow hand-dug well in a nearby creek bank.
- **Your well**—You can hand-pump small quantities of water from a well using a bought or improvised hand pump: plunge a length of pipe with a check valve on the bottom up and down in the casing, as if you were churning butter, and water will come out the top.

Longterm Storage in Small Containers

It is wise to supplement other emergency water supplies with drinking water stored in small containers.

The shelf life of emergency drinking water depends on its *original quality*, the material it is stored in, the amount of light it is exposed to, and the storage temperature. Water stored in aquifers for millions of years is fine. Water in human-made containers can be rendered unpleasant or useless for drinking by:

- the container breaking
- chemicals from the container leaching into the water
- permeation of outside chemicals through the container walls into the water
- bacterial regrowth
- algae growth

55 gal drum as roof-water catchment and storage in Panama. Much of the world lives permanently with what would be considered a state of water emergency by most readers of this book.

Well-washed glass jugs, plastic screw-top milk jugs, water bottles, etc., are good vessels for emergency water. Water stored in glass is at risk of loss via breakage. Water stored in HDPE plastic for several years is pretty much assured of having a barely tolerable plastic taste. Other plastics leach things that are nastier but tasteless. You can cover your bets by using a mixture of container materials: a bunch of glass jugs, a 55 gal *(200 L)* HDPE drum, and some polycarbonate and PETE plastic containers.

55 gal HDPE plastic drums are a classic, ubiquitous liquid storage container. They are ideal for storage of bulk emergency drinking water—when they have contained something nontoxic. Lightweight enough to carry up into an apartment empty, these drums are still movable (just barely) when full. They are tough and resilient. One 55 gal drum holds drinking water for a family of four for three weeks. Food grade steel drums are a possibility but much less desirable due to their tendency to corrode and leak.

Municipal tap water in developed countries is reliably pathogen-free and requires no treatment before storage. If you doubt your well or spring water, you can add 16 drops of chlorine per gallon before storing it. Ozonation should also work.

Label all containers with the date, water source, and method of disinfection used. Store your containers in a dark, cool place. In a few years, sample the water and see what is working for your source water, storage conditions, and taste. You can then store proportionally more water by these means.

If these filled containers are exposed to any light, they may grow algae. Even if it produces a strong taste, it shouldn't be harmful to drink.

Protecting Stored Water

In Hazards of Stored Water (p. 52) we considered what harm stored water could do. Now we're going to consider what harm can be done *to* stored water and how to avoid it. (Note: For freeze protection see p. 74.)

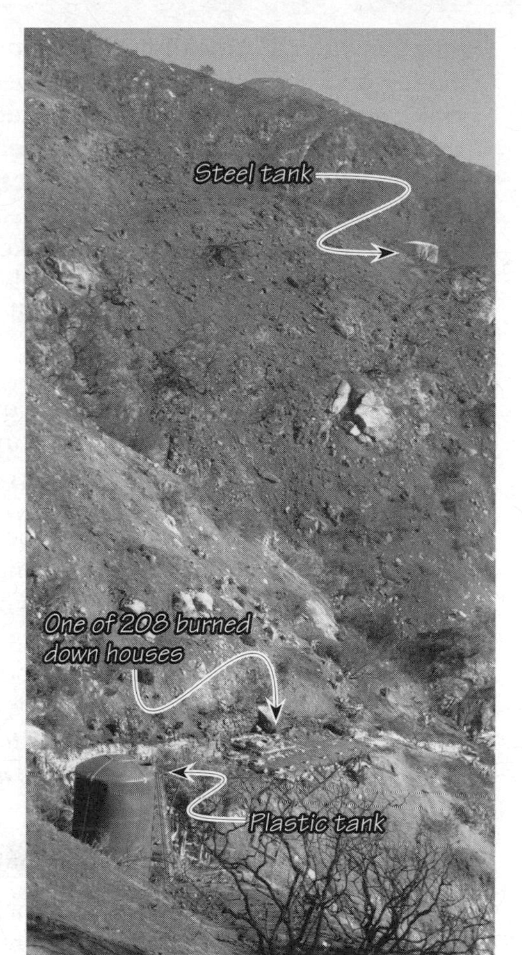

Almost all water tanks survived the 2008 Tea Fire in Santa Barbara, CA, somewhat worse for the wear, but still watertight. Several ferrocement tanks did well.

Earthquake Resistant Storage

In designing for earthquake resistance, it is important to recall that:

- inlet or outlet pipes are as or more likely to break than the tank itself
- raised storage, tower or rooftop, is much more likely to fail
- properly engineered storage is much less likely to fail

Earthquakes can generate a variety of motions—back and forth, up and down. Rhythmic movement can amplify if it coincides with the resonant frequency of the structure or sloshing water in the tank. The most violent motions the earth can generate are all but impossible to engineer structures for. Many building codes call for structures to resist an acceleration of 0.2 gravities (but accelerations of 1–2 gravities have been recorded). Imagine the plane of the earth tilting until it and the tank are sloped 20%. Would the tank fall or slide over? Perhaps it needs to be anchored.

Fire Resistant Storage

In a fire, the most likely failure points for a water storage system are the pieces that can burn. The most secure installation would not have plastic outlet pipe, rubber seals or couplings, wooden supports, or delicate steel supports surrounded by fuel, or depend on unprotected electronic controls. Water inside a tank can carry heat away, often enough to save the walls from burning. In the 1990 Painted Cave Fire—106°F, 8% humidity, 50 mph winds—400 houses burned in a few hours.[m] The sides of our old redwood water tank burned down to the water level, where the burning stopped. New water couldn't get into the tank, however, because the wooden truss work that supported the galvanized inlet line burned away, and the unsupported pipe broke.

Hurricane Resistant Storage

Tanks are vulnerable to wind when they are empty or nearly so. A full water tank, made of any material, is so heavy it is not likely to be affected by wind, except indirectly by flying debris or falling trees puncturing the side or breaking an outlet. Lightweight tanks definitely need to be anchored against high winds in areas that experience them.

Lightning Grounding

Steel tank installations without cathodic protection need to be grounded in accordance with local electrical and fire codes. Use a zinc grounding rod where the tank touches the earth, not a copper rod.

Roots and Trees

Probing, swelling roots, swaying branches, and falling trees can wreak havoc on water systems. One of the sadder but necessary maintenance tasks is to rip out tree seedlings that are too close to tanks.

Toxic Leakage or Leaching

Aquifers can be threatened by toxins from underground gasoline tanks, dry cleaners, agricultural poisons and nitrates, or saltwater intrusion driven by over pumping (see Aquifers, p. 16). **Water stored in tanks** is pretty much immune to contamination of this sort from outside; the concerns are contamination of the source water, and leaching from the tank (see How Water Quality Changes in Storage, p. 9, and Tank Materials, p. 39).

The Hazard of Permeation

Permeation is the diffusion of chemicals through the wall of the container or pipe and

Flammable Storage for Automatic Fire Control?

A neighbor used a generator to pump water from his cheap, plastic tank. One day, the generator caught fire. The fire spread to the adjacent plastic tank. When the tank burned through, the water rushed out and put out the fire on both the generator and the tank!

---

[m] ***Metric:** 41°C, 80 kph wind.*

into the water. Permeation can be an issue with aromatic toxins (gasoline, kerosene, pesticides, and the like) and plastic pipes or containers. For example, if you store emergency water in polyethylene containers (such as milk jugs) next to gasoline cans, paint thinner, and pesticides, the fumes can permeate through the plastic and contaminate the water. A municipal PVC water main passing through an industrial waste plume can absorb toxins by the same route. Permeation of toxins can be avoided by using an impervious material (metal, glass) to contain the water, keeping toxins well away, or best—keeping toxins out of your life entirely. Thick-walled plastic containers (such as 55 gal drums) are significantly less permeable than thin-walled ones.

Armed Marauders

If you are worried about hordes of barbarians stealing your water after the disaster, your best bet is to hide it underground, or disguise it (e.g., as a ferrocement boulder).

Children, Vandals, Unauthorized Access

One desert community I know occasionally found passing motorists skinny-dipping inside their potable water tanks. Fences and locks provide some security against this sort of thing. You can remove valve handles, lock valves in position, or enclose valves in locked boxes to reduce the chance of accidental or malicious adjustments.

IT Publications has this to say about children and water systems:

"Children should be considered to be compulsive saboteurs of the system. Although they do not do so deliberately, their curiosity leads to much damage and repetition of work. Open pipe ends, exposed pipeline, fresh masonry all will attract attention, with frustrating results."[16]

Systems for Firefighting

One of the most valuable uses for stored water is to prevent or limit fires, saving people and property. In places where fire safety is an issue, the firefighting performance requirements (legally mandated or owner preferred) often drive the design. Typically, the amount of storage, the pipe sizes, and pressure will be *much* higher.

Besides a hydrant and water storage set aside just for firefighting, there may be requirements for wide, gently sloped, paved access with a huge turnaround at your house. It may be cheaper to go beyond the water system requirements (even as expensive as plumbing is) in trade for slack on the road requirements, if the fire marshal is willing. From an integrated design perspective, this is almost always worth it.

Speaking of integrated design, ideas about fire safety best practices for wildland interface areas are very much in flux, stirred in part by experience in Australia and Southern California.

Climate Disruption Exacerbating Fires

A few days after the 2008 Tea Fire in Santa Barbara, California, Governor Schwarzenegger acknowledged that because of global climate disruption, Central and Southern California now have an extended fire season stretching from late February through December, instead of late June through mid-October (January and February are flood season, if we're lucky). California spent nearly a billion dollars on fire suppression in 2008, as well as half of US Forest Service firefighting dollars.

Australia had fires far worse than any in history, with larger loss of life, in February of 2009.

Fire Safety Plan

Left on its own, your home has to be *totally* impervious to fire (or lucky) to survive. The biggest factor for the survival of houses is the presence of people to defend them.

In Australia, following a particularly deadly fire season in 1983, researchers examined 100 years of fire data. More than 90% of the houses lost were never exposed to direct flame or radiant heat. They were ignited by blowing embers. Most deaths were from late evacuation. There were almost no deaths among people actively defending their homes.

In response, Australia pioneered a **"prepare, stay and defend, OR get out early"** policy in the early 1990s. Since then, more than 90% of structures defended by able-bodied people survived, because the owners were there to put out spot fires from raining embers before and after the flame front passed. Some districts in California are considering this approach. It remains to be seen how, or if, the February 2009 fires will affect this trend.

The first-person accounts from fire disasters are harrowing and informative.

In Santa Barbara's Tea Fire, 800 students sheltered safely in the windowless, concrete gym in a well-cleared area of Westmont College. This was planned in advance; they'd had a drill less than month before. There were no special provisions for air supply.

In Australia, one family survived the recent inferno in a homebrew fire bunker attached to a water tank. The $1,000 fireproof door heated up so much before they got in that it wouldn't shut all the way; flames licked in. There was no mention of special provision for air supply.

Five friends had seconds' notice before their shop and vehicles were engulfed in flame. They jumped into waist-deep water in an old concrete water tank. They stuffed wet shirts in the cracks in the walls, took turns boosting each other up to the hatch for (smoky) outside air, and survived.

Some people evacuated early, but miles away their escape was cut off by another fire. Some stayed, fought, and didn't make it.

Our home has been threatened by fire three times in the past two years. To be willing to stay (or if we're trapped), I'd want at least one totally fireproof structure we could hide out in for the 5–30 minutes it takes the flame front to pass.

There are unanswered questions about thermal resistance, thermal mass, insulation, and air supply. There does not appear to have been much research or testing. Analysis of the tragedy unfolding in Australia as we go to press will likely help others survive future conflagrations.

Cob (monolithic adobe fibercomposite) is a material which is more fire-resistant than concrete, because it conducts heat more slowly. It is also earthquake and termite resistant, and exceptionally owner-builder friendly. A cob cottage with a ferrocement roof and metal shutters over metal dual-glazed windows and doors is a structure which could potentially resist the most intense firestorm, and still be pleasant to live in. Small cob cottages are reportedly used traditionally to safeguard family treasures in Japanese wooden villages.

Water tanks seem to survive fire very well compared to other structures. Could a ferrocement rainwater harvesting tank near the house do triple duty as an invincible (if claustrophobic) fire-safe haven? It seems that a higher roof to ensure head space with air inside the tank, and a pressure tank with fresh air supply, could make it so. The cost would be very low and security high compared to any alternative.

Buried, cast concrete septic tanks as low cost, high security underground fire bunkers are being promoted in Australia (probably better to have it double as a root cellar rather than a septic tank). In the old days, people in Australian lumber camps had dugouts with timber roofs as fire shelters, fed by the inches of fresh air rushing along the surface of the earth to feed the fire.

Lots of people living in flammable brush is inherently dangerous. The least dangerous way of managing this is an open question.

Whatever your fire safety plan, research it well, prepare, and practice for it; the stakes are quite high.

Current Southern California Wildland Interface Fire Safety Approach:

- *Mechanical clearing around flammable homes and landscapes*
- *Big, wide, costly, and ecologically devastating evacuation/access roads*
- *Water systems that often fail in a fire*
- *Expensive but flammable construction*
- *Protection of homes only by overwhelmed professionals*
- *Dense fuel that has frequently burned clear the past 15,000 years, but has built up unprecedented levels from 80 years of fire suppression*
- *A state insurance system for high fire areas that may go bankrupt due to climate change-induced increase in fire, even though it currently is paying out only half the cost of rebuilding*

The more economical and ecological approach I think we may eventually be forced to adopt:

- *Fire-safe landscaping, most clearing by frequent, less intense fires*
- *Human-scale roads*
- *More robust water systems*
- *A mix of structures that are passively and actively fire-armored to protect residents in a full scale firestorm, and inexpensive structures that burn clean*
- *Protection of structures primarily by residents*
- *Maintenance of low fuel levels through frequent fires*
- *Increasing reliance on self-insurance*

—oasisdesign.net/shelter/fire

Water System Design For Fire

In most fire emergency situations, well power is off, and/or the flow demand is so much greater than the supply that storage is essential to cover it. For example, the incoming supply might be 10 gpm, while the fire department can go through several thousand gallons in 15 minutes.[m] However, incoming water may make a crucial difference in reality—especially if the reality is that your tanks are low to start with, or get drained wetting things down before the fire even gets there.

Armoring the water supply system, as described in Fire-Resistant Storage (p. 77), will increase the likelihood that at least *some* water is being added to the system even as you are rapidly draining it. **If you don't have an entirely gravity powered system, stored water may be your only supply,** as fire often knocks out the power grid before the

[m]***Metric:*** *Incoming supply is 40 lpm, fire department can use tens of thousands of liters in 15 minutes.*

flames arrive. A battery bank and electric pump, a gas pump, or a generator and electric pump can keep water flowing/pressurized in this instance (don't plan on your generator or gas pump working in an emergency unless you maintain it).

In general, I prefer fire emergency hardware, especially pumps, that are incorporated into the regular system. I plan to use the fire emergency pump to pressurize the rainwater supply to our house. Besides the efficiency of the item doing double duty, it's much more likely to work when a fire comes if it is something that is regularly used.

When plumbing your pool or spa, hook things up so that you can use the pool's own filter pump to power a fire hose. Some people also include a gas pump (or generator) in the pool plumbing for firefighting. This may require different-sized lines and/or pump for adequate flow and pressure. Make sure the pump inlet is near the bottom of the pool, so the system can access most of the water. If you use a gas pump for irrigation, you might as well use it for fire, too.

Water elements for fire safety (besides the idea of using a water tank as fire-safe haven, mentioned earlier) are of three classes: systems to support fire hoses, automatic fire sprinklers (interior and/or exterior), and water to refill fire trucks:

Water for Fire Hoses

To operate fire hoses, you will need:

- **Stored water with high pressure (gravity preferred)**—At the hydrant, pressure should be 40–100 psi *(275–700 kPa)*.
- **A decent-sized line**—For good dynamic pressure at high flow. For instance, a high-pressure 2" line can barely supply two 1-½" hoses at the same time, while a 3" line at the same pressure could supply three hoses, with better flow for each hose.*
- **Fire hoses**—Stored onsite, convenient to hydrants.
- **A pump**—To make up for low pressure and/or an inadequately sized line.
- **A foam injection system**—See Foam, p. 81.

Even one fire hose places an extreme demand on system hardware. Accommodating this can double the resources required to build your storage and distribution plumbing—all for something that hopefully will never get used. Is it worth it? It is a form of insurance, one that you'll have to judge how much to purchase.

The difference between a ¾" garden hose and a 1-½" fire hose is truly phenomenal. Likewise the difference between 40 and 100 psi of water pressure. You'll need a powerful water delivery capability to have a chance against the considerable power of a house fire or wildfire. When we are burning brush piles, occasionally a gust of wind drives the flames skyward on a windrow of 50 truckloads of tinder-dry brush. In seconds it can go from feeling like a nice campfire to having your clothes ironed with you in them. A moment's burst from a 1-½" fire hose (with 60 psi of pressure) puts it right down. A garden hose going full-blast would do essentially nothing; it would just take a bit longer for it to get so hot that you had to back off. If we were to let the whole thing get fully engaged and roaring, with wall-to-wall two-story-high flames, we could probably put it completely out in a few minutes with two fire hoses.

On the other hand, a fully engaged firestorm in Southern California chaparral, whipped by freeway-speed, hot, dry wind is beyond the capability of *any* fire hose to suppress. That's when you 1) pay your insurance diligently and get out early or 2) build a fireproof bunker and stay and defend against embers (p. 79).

Some other things to think about are the location of fire hose standpipes (small hydrants) and how they are plumbed. The hydrants should be near structures, but not so near that it would be too hot to hook up hoses if the structure were burning.

The plumbing needs to be secure. I've seen hydrants that were made by connecting a hip-high vertical length of steel pipe directly to a tee in an underground PVC pipe. Imagine a panicked person pulling hard for a bit more hose they desperately need to keep their home from burning down. Pulling, that is, on a long steel lever with a brittle connection to a delicate plastic pipe. If it breaks, they've got a tough decision—turn off the water, or watch the tank empty uselessly, in each case while the house burns.

FIGURE 27: SWING JOINT ATTACHMENT FOR A SMALL HYDRANT

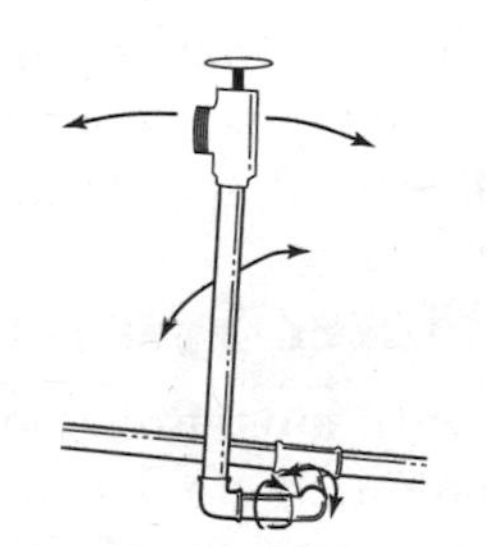

Hydrant can be pulled or knocked over without breaking the main. Use plenty of Teflon paste on the threads and wrap the outside with pipe wrap tape.

**Flow is proportional to pipe diameter squared times two thirds; 230′ (70 m) of head provides 100 psi (700 kPa), ideal for fire hoses.*

One solution is to encase the steel in enough concrete at ground level so that the steel will bend before the plastic underneath breaks. A better solution is to put two 90° bends at right angles, so the standpipe can be pulled or knocked over without stressing the plastic pipe (Figure 27).

Water for Fire Sprinklers

Water system features for fire sprinklers:

- Stored water with adequate gravity pressure (or a booster pump), and a line size engineered for adequate flow. (¾" at 100 psi to 1-½" at 40 psi for a residence.)[m]
- A mechanism for triggering the sprinklers at the right time, such as heat-sensitive sprinkler heads. I've also heard of heat-sensitive wires that trigger a valve.
- Roof-wetting systems on a home with a roof rainwater harvesting system can be plumbed to recycle the water running down the gutters, so it takes much longer to run the tank dry.

This house was brick with a tile roof. The owner had his own gravity flow water supply feeding a 1 ½" fire hose. He stayed and sprayed down the exposed rafter tails with fire retardant gel until his hair was singed from the heat of the advancing wall of flame.
Despite his doing everything right, the house burned anyway, probably from the underside of the roll roofing under the tiles on the eaves catching fire. Cars sprayed with foam survived untouched.

Indoor fire sprinklers may or may not stop a house fire, but they will virtually always slow and cool it enough so residents can escape. [33] There is a detailed code for interior fire sprinklers.

For exterior sprinklers—designed to keep a wildfire from incinerating your home—the design is up to you. The common roof-wetting design is a high-pressure irrigation sprinkler or two.

Here's my (untried) design idea for rooftop sprinklers: Place copper pipes along the roof ridgelines, with small holes drilled in them for the water to jet (or drool) over the roof. It seems that the water would get used much more efficiently, and that you'd cover the whole roof even if the pressure were really low (as it will be if your neighbors are all wetting things down in a panic). What's more, if you've got a roof rainwater harvesting system, the water would virtually all go back into your tank, so you could just turn the pump on and leave the water to circulate. The same type of system (pipe with holes) could be installed under the eaves to protect eaves and windows as well, although it wouldn't recycle. A flowing sheet of water over tempered, dual glazed windows is an effective fire wall (if you try one of these, let me know how it turns out).

Could "stay and defend" have saved this house? Having a fire-safe bunker would have enabled the owner to find out with an acceptable level of risk (and could have saved his neighbors months in the burn unit).
Note that the half-full plastic tank survived the fast moving firestorm.

Water for Fire Trucks

Fire trucks don't carry that much water—usually no more than 800 gal *(3 m³)*. Dedicated tanker trucks can carry a few thousand gallons *(10–15 m³)*. Sources for refilling trucks include:

- stored water with gravity pressure, a large diameter line, and the right hydrant fitting (a 4" hydrant right by the tank is common)
- a pond, swimming pool, hot tub, or river, pumped out with a suction hose

Foam

Injecting Type A firefighting foam or gel greatly increases the effectiveness of water for fighting fires. The wet foam sticks, smothering fire on something burning, or insulating and reflecting heat from something you're trying to protect.

If your plan is to wet things down and evacuate early, foam is much more effective than water alone. Instead of running off, it sticks (for 20 minutes to a few hours). Turn your house into a big marshmallow and go (or duck into your fire bunker).

A neighbor of ours did this with his house and big wooden deck. He watched the flames cavitate under the deck, but it didn't catch fire—quite impressive.

---

[m]***Metric:** Line sizes range from 2 cm at 700 kPa to 4 cm at 275 kPa.*

Chapter 7: Examples of Water Storage Systems for Different Contexts

Poor Surface Water Quality, Limited Groundwater

Context: A community of 100 in arid Southern California desert.

Goals: Provide for conservative residential use for 30 homes, with a maximum use of 300 gpd *(1.1 m³/day)* per household. Provide substantial fire safety reserve.
For the first 70 years, economy and owner-serviceability were primary goals. Owners have recently, reluctantly decided to trade off economy, owner-serviceability, and environmental impact in order to meet the requirements of the Surface Water Treatment Rule[34,35] and minimize possible liability exposure.

Water supply: Gravity flow springs (5–17 gpm/*20–60 lpm* diverted from 8–30 gpm/*30–115 lpm* flow) with turbidity and legal challenges, well from an aquifer with nitrate levels hovering just below the legal limit (14 gpm/*53 lpm*, 12 hrs/day max pumping); some harvesting of rooftop rainwater and runoff.

FIGURE 28: CALIFORNIA DESERT WATER SYSTEM ELEVATIONS

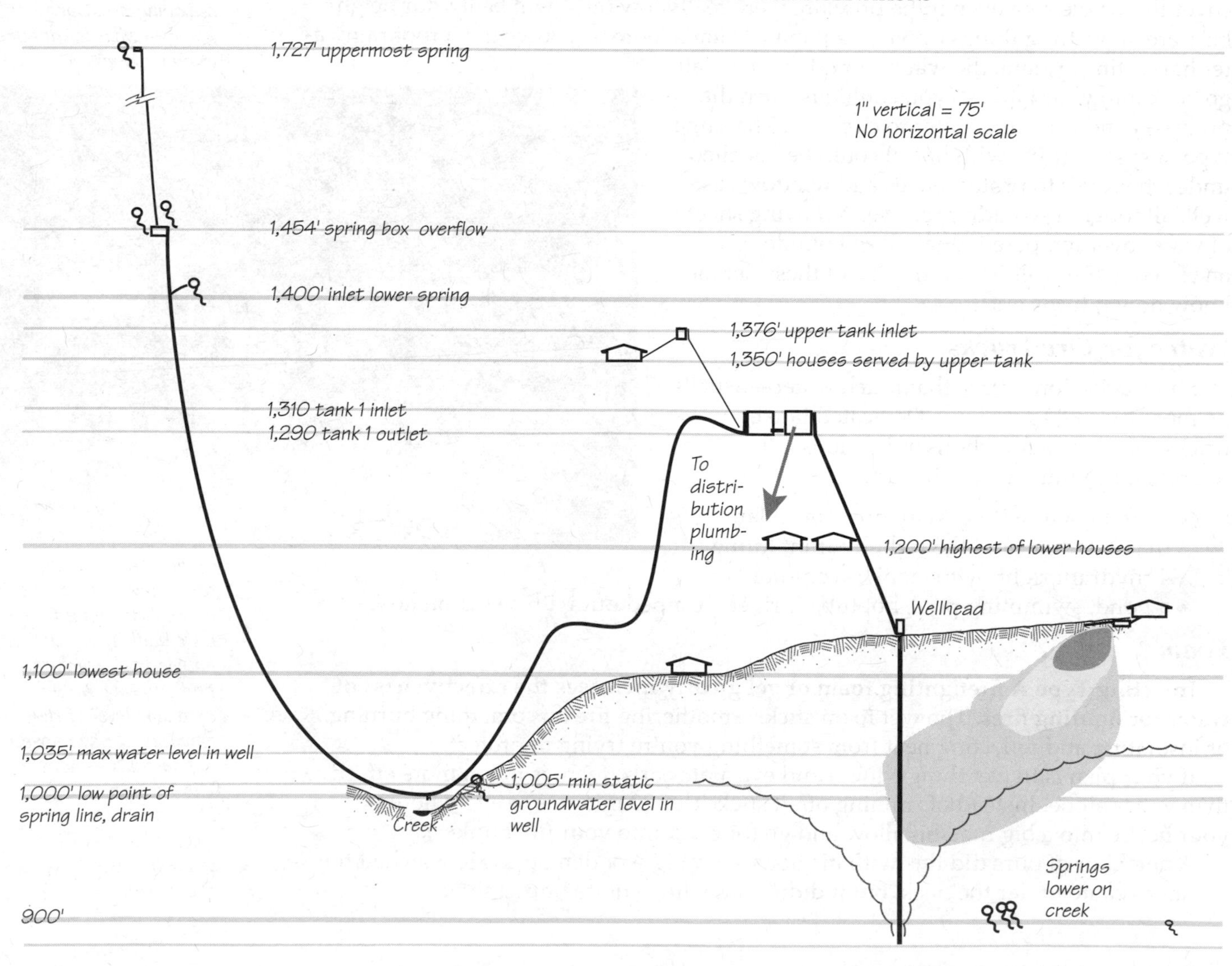

Storage: Small aquifer (with a capacity of approximately one or two years' consumption), three tanks (two 50,000 gal/*190 m³* tanks, one 3,000 gal/*11 m³* tank), about 10,000 gal *(38 m³)* in three private rainwater harvesting tanks.
Average use: 4,500 gpd.
Peak use: 9,000 gpd for up to 20 consecutive days of really hot weather.
Fire design flow: System can deliver 100,000 gal *(380 m³)* at 35 psi *(240 kPa)* to a 6" hydrant. It can also power one or two 1-½" fire hoses on the distribution network, which is 2" pipe, with pressures from 38 to 100 psi depending on elevation.
There are about 40,000 gal *(150 m³)* of emergency set-aside that can only be accessed by opening a valve at the tanks. This is not ideal; it should also be directly connected to the hydrants.
Water security: Good. There is sufficient storage to meet two weeks' use without water income. In the absence of electricity, this system can deliver clean (if not legally compliant) spring water equal to half the peak dry season consumption. In the case of flood waters contaminating the springs or washing out the pipes, it can deliver well water equal to the full rainy season consumption.
There is concern about falling water tables due to overdraft all around, but this has not hit the springs nor this well directly as yet. (Except that the groundwater level was lowered due to overdrafting the well during the five years when the springs were offline while code-compliant treatment was being installed.)
Issues and notes: This system is above the ten-connection threshold and is thus subject to the draconian provisions of the Surface Water Quality Act. The spring water is substantially pre-treated in the tank first. With settling and ozonation, the system delivers coliform-free water with <1 NTU turbidity at up to 5 gpm, which meets the sky-high performance requirement of the law. Nonetheless, a package treatment plant, an online turbidimeter, a chlorine injector, and an online confirmation of chlorination meter are being added to comply with the *proscriptive* requirements of the law. This law is a classic example of tunnel vision (see p. 5). Hopefully the law will be made more flexible in the future. Allowing users to meet the performance *or* proscriptive requirement (as is the case with building codes) would be a great improvement.

Only Stored Water in Dry Season, Hydroelectric in Wet Season

Context: Ecovillage and events center of 30 residents in Highland Central Mexico.
Goals: Provide storage for highly conservative residential use for ten homes, with a dry season max use of 106 gal *(0.4 m³)* per day per household. Low cost, low impact, and simplicity are paramount. Water quality goal is <10 fecal coliforms per 100 ml.[4] Fire safety would be great but is beyond the means of the owners.
Water supply: Rainfall and runoff, during four- to six-month rainy season only. During the dry season, all water used is drawn from storage, and the storage level drops each day. There is a community water diversion at the base of a waterfall that runs primarily during rain, at zero to 265 gpm *(1 m³/min)*. Every home harvests its rooftop rainwater. In a pinch, people can buy tanker trucks of water driven up from the valley 500 m below.
Storage: There is a rock dam spanning two vertical veins of bedrock, which forms a community tank of 100,000 gal *(375 m³)*. A smaller rock dam forms the waterfall diversion settling pool of 11,600 gal *(44 m³)*. There is a ferrocement distribution tank of 2,600 gal *(10 m³)* used to meter the water (by the tankful) as it is transferred from the big community tank to about 20 private tanks of ferrocement, rock, cement, or plastic, which altogether hold another 53,000 gal *(200 m³)*.
Average use: 600 gpd *(2,256 lpd)*.
Peak use: The rate is unknown, but peak flow occurs during big festivals hosted at the site, with hundreds of attendees.

Water security: Poor. In the absence of electricity, many homes could not get water out of their buried cisterns easily. If the monsoon is late, or there is a leak, individual homes and sometimes the whole community run out of water and have to have it trucked in.

Issues and notes: Huehuecoyotl is a critical context for water storage.

They have zero water income for six to eight months of dry season. No rain, no creek, no springs, not even reachable groundwater. (The World Bank dug a 400' *(120 m)* deep well at the village next door, and it was dry as dust at the bottom.) All their water during the dry season is from stored water, of which there is less each day.

Then, in the monsoon it rains 4.5' *(1.5 m)* in four or six months!

One interesting thing about this place is that the 11,600 gal *(44 m³)* pool at the base of the waterfall can be used as a hydroelectric storage "battery" during the rainy season. That is, draining the pool through the hydroelectric turbine yields electricity, just like a battery.*

There is more on the rainwater harvesting and seasonal hydroelectric systems at Huehuecoyotl in our forthcoming book, *Rainwater Harvesting and Runoff Management.*[3] The greywater systems are described in *Create an Oasis with Greywater*[2] and *Branched Drain Greywater Systems.*[36]

The "presa pyramida" (pyramid dam)—a 100,000 gal (380 m³) cistern made by sealing off the space between two walls of bedrock with stone and cement. This construction is difficult to seal perfectly leak-tight, which is important if you don't have water to add to the cistern for eight months.

Table 8: Huehuecoyotl System Components Sample Portion

Huehuecoyotl System Components, Valves, and Settings

Location code-Functional code	Functional description of component	Height (m)	Type	Size	Material	Dry season	Rain start	Rainy season	Clean	Fill	Hydro	Note
	Description					Standard states						
Tinaco Huehue East			Tank	50 m³	FC	Full/MT	MT	Full	MT		Full	
HD-TE	Inlet pipe	2,012.25		2"	pvc							
TE-IS	Inlet shutoff	2,012.00	ball valve	2"	pvc	shut/open		shut/open		**open**		OK to leave open when PDLC is pressurizing line if water is clean, especially if the tinaco is basically full and it is just topping off.
TE-IF	Inlet float		float valve	2"	brass			open/shut				
TE-F	Outlet filter		mosquito net									
TE-OL	Outlet level control		pipe	1.5"	galv	15 cm below H20		well below H20	**high**			Remove union and unscrew to get last few m³ of water
TE-OL	Outlet	2,009.00	ball	2"	brass	open	open	open	**shut**			
TE-D	Drain	2,008.80	threaded cap	2"	galvanized	shut	shut	shut	**open**	shut		
TE-ISR	Rain water inlet shutoff	2,012.30	rubber stopper	1.5	rubber	shut	shut	open	**shut**	**open**		
	Rainwater inlet filter		*mosquito net*									
TE-RB	Rainwater bypass	2,012.30	plug	2"	pvc	open	open	**shut**				Sends water inside tank when shut
TE-	Outlet line			1.5"	pvc							
Meter Tees												
P	Drain plug		threaded plug	1.5"	galv				**open**			In path-open to empty tinaco E,
S	Shut-off to theater, 65 m³ cistern, bathhouse		gate	1"	brass	shut	shut	shut	shut	**open**		shut for leaks, service
M	Meter to theater, 65 m³ cistern, bath house			1"	brass			-				
S	Shut-off to house		gate	1"	brass	open		open	**shut**			shut for leaks, service
M	Meter to house	2,001.00		1"	brass			-				

Code key			
IS	Inlet shut-off	SA	Shock absorber
IF	Inlet float valve	VB	Vacuum breaker
OS	Outlet shut-off	M	Meter
FO	Outlet filter	PU	pump
OL	Outlet level control	R	Rain
O	Overflow	L	Line
AR	Air relief	B	Bypass

**11,600 gal with 70' of head (44 m³/21m) run through a Harris microhydro turbine provides about 1,800 watt-hours of energy, about the same as a big deep cycle battery, but with much longer life and no problem with 100% cycling.*

Creek Direct with Remote Storage and Sand Filtration

Context: Rural conference center on 30 acres in Southern Oregon.

Goals: Provide for conservative residential use for up to 30 seasonal workshop participants and a handful of year-round resident caretakers. Provide irrigation for ½ acre *(2,000 m²)* of gardens and orchard. Fire safety is beyond the scope of this system.

Water supply: Year-round creek (>30 gpm/*110 lpm* flow).

Storage: System runs creek direct, with a 300 gal *(1.1 m³)* tank at the far end of the system. This is intended to cover peak demand flow, supply more secure water to the highest, last houses on the line, and provide reserve for when the diversion washes out and the system needs air flushed from the lines.

Average use: 600 gpd *(2.2 m³)*.

Peak use: 900 gpd *(3.3 m³)* for up to 20 consecutive days of really hot weather during workshops.

Water security: Good. This system does not use electricity at all. In the case of flooding, the water is not drinkable, and the diversion washes out. However, this only happens in winter, when the population is low, and the 100 gal *(0.3 m³)* of reserve is sufficient to hold over residents until the diversion can be restored.

FIGURE 29: CREEK DIRECT WITH REMOTE STORAGE SYSTEM

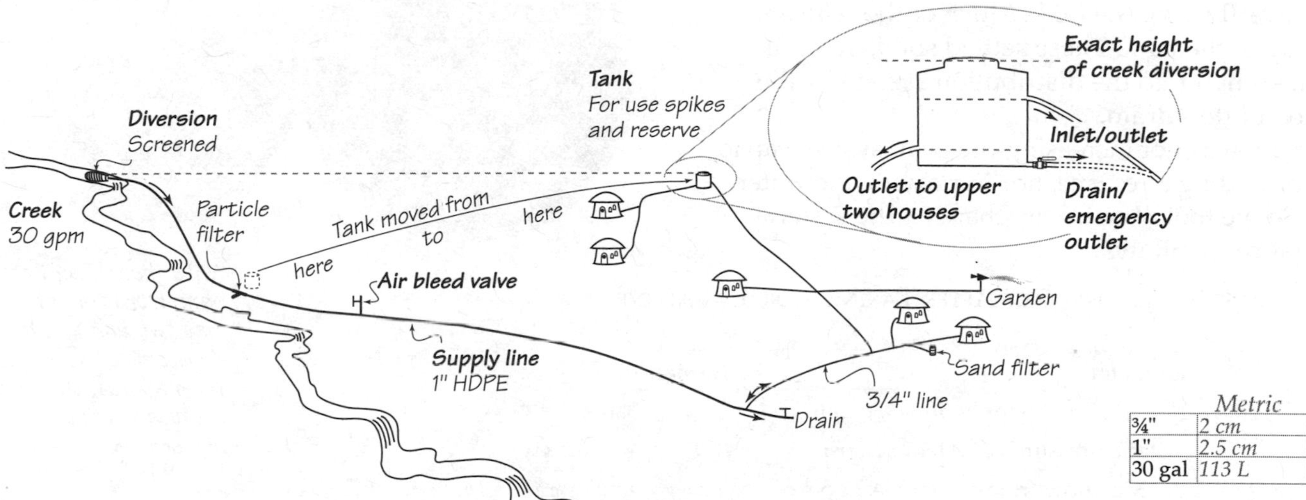

	Metric
¾"	2 cm
1"	2.5 cm
30 gal	113 L

Issues and notes:

- The creek water, one branch of which comes out of a recent clear cut, had higher than desirable coliform bacteria.
- The system was providing less water flow than desired.
- When the water line washed out, someone had to go out for hours (usually in the dark, during freezing rain) and de-couple the line in several places to let air out before the flow would start again.
- The tank was defiling the most beautiful place on the creek, and the owners wanted to move it.
- Finally, the highest two buildings on the system—the last on the line—didn't have as much water security as they'd like; any open valve lower down would leave them high and dry.

First, the bacteria issue. We inventoried all the possible drinking water sources, which included a very small, shallow spring nearby, bigger and deeper but distant springs, "naturally" filtered creek water coming out the bottom of an old dam filled with sediment, and the creek itself. We concluded that no other source was worth making a separate delivery system for, and that the best option was to treat the creek water itself with a slow sand filter (Figure 30, at right).

FIGURE 30: SMALL SAND FILTER FOR DRINKING ONLY

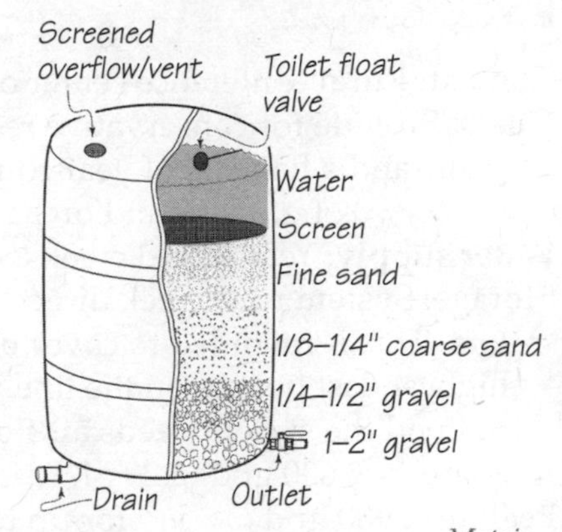

	Metric
1/8"–1/4"	.3–6 cm
1/4"–1/2"	.6–1.27 cm
1–2"	2.5–5 cm

Now we're going to look at how a minor change in the storage for this system did a lot to resolve the other issues:

ORIGINAL SYSTEM PERFORMANCE

Parameter	Performance	
Supply flow	30 gpm	*113 lpm*
Static pressure in system	12 psi	*83 kPa*
Max flow from system	3 gpm	*11 lpm*
Reserve on system failure	0 gal	*0 L*
Time to restart system	4 hours	
Daily use	600 gal	*2 m³*
Water contributing to sediment load, daily	20,000 gal	*76 m³*

The water source, a creek, flows several times the maximum flow capacity of the water line. Of the 30 gpm *(110 lpm)* of creek flow, a 1" pipe captures about ⅔ (20 gpm *(75 lpm)*) and diverts it into the tank. Of that, only about 1 gpm *(4 lpm)* average is consumed, with the balance overflowing the tank. Much of the settleable solids stay in the tank. These settled solids were likely to be vacuumed into the distribution system by the combined outlet / drain.

The tank was not increasing the peak system capacity, not providing a reserve, nor improving the water quality. So we took it out. This changed the system performance as follows:

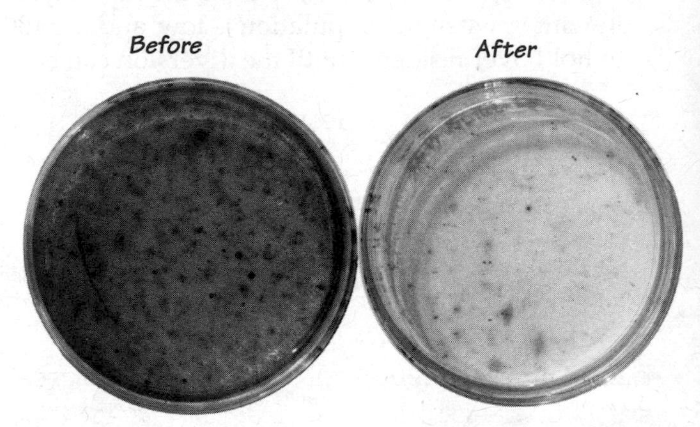

Water before and after sand filtration, after just two and a half days online. As the filter matures, the treatment level will be much higher still.

SYSTEM PERFORMANCE AFTER TAKING STORAGE OFFLINE

Parameter	Performance		
Supply flow	30 gpm	*113 lpm*	Same
Static pressure in system	17 psi	*120 kPa*	40% more
Max flow from system	4.2 gpm	*16 lpm*	40% more
Reserve on system failure	0 gal	*0 L*	Same
Time to restart system	2 hours		50% less
Daily use	600 gal	*2 m³*	Same
Water contributing to sediment load, daily	600 gal	*2 m³*	97% less

Then the caretaker had an inspired idea for what to do with the tank—put it at the *far* end of the distribution system. Probably only someone with little water system design experience could have conceived of such an unconventional geometry—usually the storage is at the beginning of the distribution system.

I suggested putting the tank at a level such that the overflow is 2" *(5 cm)* lower than the creek diversion several hundred pipe feet *(a few hundred meters)* away on the other side of the ridge (see Figure 29, previous page). With the tank inlet / outlet located a third of the way down, the tank fills whenever all the valves down below are shut. When someone turns on a garden sprinkler while someone is in the shower, water flows both from the creek diversion and from the upper third of the tank, giving the bather a good chance of getting the soap off before the flow drops.

By adding a separate outlet two-thirds of the way down, the tank provides water pressure and supply for the upper two houses that is not affected by use in the rest of the complex.

With an outlet for the reserve at the bottom of the tank, there is emergency water that can run the whole place for a couple days if the line washes out, or so that the maintenance can be done at a convenient time. Also, the pressure from the reserve water can be used to push the air out of the system so that it starts up with far less effort. The total cost of all the changes to the system is about $40 for two bulkhead fittings to make new outlets on the tank, and an inline particle filter to ensure that solids from the creek don't fill the line. The following chart shows the system performance at the end, as compared to the beginning—quite an illustration of how the design of storage can totally change the performance of a system:

SYSTEM PERFORMANCE AFTER MOVING TANK TO FAR END OF DISTRIBUTION SYSTEM

Parameter	Performance		
Supply flow	30 gpm	*113 lpm*	Same
Static pressure in system	17 psi	*120 kPa*	40% more
Max flow from system	4.2 gpm continuous/8 gpm for 15 min	*16 lpm/30 lpm*	260% more
Reserve on system failure	300 gal	*1 m³*	300 gal more
Time to restart system	½ hour		75% less
Daily use	600 gal	*2 m³*	Same
Water contributing to sediment load, daily	600 gal	*2 m³*	97% less

Very, Very Low Pressure

Context: Single-family residence in a steep, wild canyon in Southern California.

Goals: Provide for conservative residential use for one family of four, by gravity. Provide irrigation for 1/8 acre *(500 m²)* of gardens and orchard. Fire safety is beyond the scope of this system.

Water supply: Horizontal hard rock well/infiltration galley (a tunnel with water percolating in or out of it), 1–2 gpm flow *(4–8 lpm)*.

Storage: With the kitchen tap shut, the shallow pool in the floor of the infiltration galley rises until it meets the overflow. With the tap open, the pool level drops as much as 6" *(15 cm)*, to the spill point of the outlet.

Average use: 100 gpd *(0.4 m³/day)*.

Peak use: 200 gpd *(0.8 m³/day)*.

FIGURE 31: ULTRA LOW HEAD STORAGE

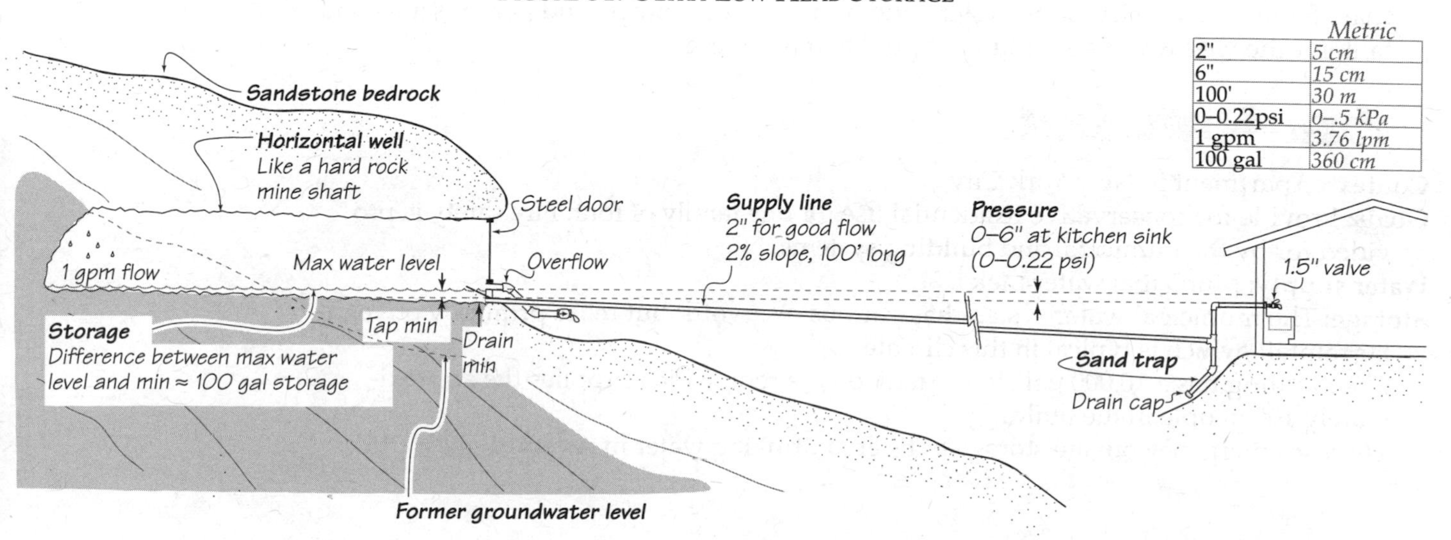

Water security: Excellent. This system does not require electricity and has only a short run of pipe. Floods affect neither the pipe nor the well. It is thus quite secure. Falling groundwater is about the only thing that could threaten this water supply.
Issues and notes: This system runs off of almost no pressure, yet delivers several gpm.

Simple Jungle Eden

Context: A community of 20, 45 minutes' walk into the jungle from the Caribbean coast of Costa Rica.
Goals: Provide for conservative residential use for five families. Irrigation is all directly by rainfall. Fire safety is not an issue.
Water supply: Creek of about 5–30 gpm *(20–110 lpm)* is used directly for drinking, clothes washing, bathing (no soap). Small amounts of water are hand-carried into the homes.
Storage: The storage is in the soil and aquifers that supply the creek, plus a few gallons in small containers in the homes.
Average use: 100 gpd *(0.4 m^3/day)*.
Peak use: 200 gpd *(0.8 m^3/day)*.
Water security: OK. This system is only affected by extensive runoff during high water, lowering the quality.
Issues and notes: This system has no artificial plumbing at all.

Rural House with Well

Context: Single-family residence in the Arizona desert.
Goals: Provide for conservative residential use for one family of four, plus irrigation of ½ acre. Wildfire safety is not a major issue due to low fuel load in the surrounding desert.
Water supply: 400' *(120 m)* well, with safe yield of 200-1000 gpd *(0.8–3.8 m^3/day)*. Water level varies between 190' and 230' *(60–70 m)* below the surface. The tank is 90' *(30 m)* above that.
Storage: The storage is in the aquifer that supplies the well, and in one 2000 gal *(7.6 m^3)* tank.
Average use: 200 gpd *(0.8 m^3/day)*.
Peak use: 400 gpd *(1.6 m^3/day)*.
Water security: Poor. This system is critically dependent on electricity. Without electricity to power the pump, there will be no water in five to ten days.
The aquifer is being lowered by overdraft from neighboring wells, and at some point the well could go dry.
Issues and notes: I would feel better about the water security in this spot with the addition of a rainwater harvesting system and its associated storage, and perhaps a second tank for the well water also, plumbed so that it is all reserve.

Urban Apartment

Context: Apartment in New York City.
Goals: Provide for conservative residential use for one family of four. Fire safety is provided for by the municipal and building systems.
Water supply: Municipal water meter.
Storage: The municipal water system has various reservoirs, but relies primarily on regular rainfall, which is typical in this climate.
The building has a 10,000 gal *(38 m^3)* tank on the roof, which supplies the approximately 100 people in the building.
There is emergency on-site storage of filtered drinking water in several 1 gal HDPE

screw-top milk jugs, glass apple juice jugs, a 5 gal polycarbonate jug, plus 30 gal in the hot water heater and 4 gal in the toilet tank. If push came to shove, there are 30 gal in a freshwater aquarium.

Average use: 150 gpd *(0.6 m³/day)*.

Peak use: 300 gpd *(1.1 m³/day)*.

Water security: OK. This system is critically dependent on electricity. There is a small rooftop tank, but without electricity, there will be no water in a few hours at most.

Issues and notes: I would feel better about the water security in this spot with the addition of a 55 gal drum of clean stored water.

Swank Suburban House

Context: A 5,000 ft² *(460 m²)* single-family home in a new gated community on the outskirts of Los Angeles, at the wildland/urban interface. Home is bordered by mountainous dry scrub on upwind side.

Goals: Provide for extravagant residential use for one family of four. Supplement fire safety provided for by the municipal system to the extent practical.

Water supply: Municipal water meter, with water from the Colorado River and Owens Valley.

Storage: The municipal water system has various reservoirs, but the primary water sources are hundreds of miles away. The local reservoirs require electricity to get to this location and are of questionable earthquake-hardiness. There is hardly enough local water to meet residents' drinking needs.

The home has a 30,000 gal *(113 m³)* swimming pool.

There is emergency on-site storage of filtered drinking water in several 1 gal jugs, plus 50 gal in the hot water heater and 16 gal in the four toilet tanks.

Average use: 1,000 gpd *(3.8 m³/day)*.

Peak use: 2,000 gpd *(7.6 m³/day)*.

Water security: Poor. This system is critically dependent on electricity, and a very long, weak supply line. Without electricity, there is no water immediately. The greatest vulnerability, however, is of a large earthquake damaging major portions of the supply system. At this location, people could potentially be entirely without water for months. If the earthquake precipitated an economic crisis, this house could be without water indefinitely.

Issues and notes: The pool does quite a lot to improve water security at this spot. It could be further improved with the addition of a few 55 gal drums of drinking water. Fire safety would be improved with the addition of a pump and fire sprinklers and/or hoses.

Appendix A: Measurements and Conversions

How units are dealt with in this book:

Any measurement clearly expressible without numbers or units is expressed without them (e.g., "an arm's length").

Where the text flow is too chopped up by non-essential numbers, they are relegated to footnotes. I skipped metric conversions of pipe sizes in the text. They are all here.

Measurements for examples or construction plans are expressed in the national units where I think most tanks of this type will be made.

If the units are approximations or don't really matter, the units are given to the nearest round number. Thus, a photo of a tank might be captioned "10,000 gal *(40 m³)* tank" rather than "10,000 gal *(37.854 m³)* tank."

Using the tables:

Everything in the same row is equal. For example:

1'=12"=0.3 m=30 cm.

Length / Height

Feet ft'	Inches in"	Meters m	Centimeters cm	
1	12	0.30	30	Long as… your foot
0.08	1	0.025	2.5	Long as 1st knuckle of the thumb
3.28	39.37	1	100	A long stride

Area

Square meters m²	Square feet ft²	Acres ac	Hectares ha
1	10.76		
0.09	1		
4,047	43,560	1	0.405
10,000	107,639	2.47	1

Pipe Sizes

US	Metric
½"	15 mm
¾"	20 mm
1"	25 mm
1-¼"	32 mm
1-½"	40 mm
2"	50 mm
2-½"	65 mm
3"	80 mm
4"	100 mm
6"	150 mm
12"	300 mm

Volume

Cubic meters m³	Liters L	Gallons gal	Cubic foot ft³	
1	1,000	264.17	35.3	A cube 40" on a side
0.001	1	0.26	0.0353	A liter is slightly more than a quart
0.00387	3.78	1	0.134	1 US gal = 0.833 Imperial gal
0.0283	28.3	7.48	1	
1,233	1,233,482	325,851	43,560	1 **acre foot** (af)
0.2	200	55	7.06	Big plastic or steel drum
113	113,562	30,000	4,010	Standard suburban swimming pool
0.00235	2.35	0.62	0.0829	1" of rain on 1 ft²
	1			1 mm of rain on 1 m²

Pressure

Abbreviations: Atmospheres (atm), Pounds per square inch (psi), Kilopascals (kPa)

atm	psi	kPa	Feet	Meters	
1	14.70	101	33.9	10.3	A 1"x1" column of air from the Earth to space in height weighs 14.7 lbs
0.0680	1	6.89	2.31	0.703	Each 2.31' adds 1 psi of pressure
0.0099	0.145	1	0.334	0.102	
0.0295	0.433	2.99	1	0.305	Each foot adds 0.433 psi of pressure
0.0978	1.42	9.80	3.28	1	Each meter adds about 10 kPa
6.80	100	689	231	70.3	Max pressure for household plumbing, ideal static pressure for fire hoses
1.70	25	172	57.7	17.6	Min. pressure for washer, demand heater valves

Flow

An industrialized world household of four uses roughly a 1 m³ a day of water.

A non-industrialized world household of ten uses roughly 1 m³of day.

Abbreviations: Gallons per minute (gpm), Cubic feet per second (ft³/sec), Liters per minute (lpm), Cubic meters per day (m³/day), Acre-feet per year (afy)

gpm	ft³/sec	lpm	m³/day	afy
1	0.00222	3.79	5.45	1.61
448	1	1699	2,446	723
0.26	0.000589	1	1.44	0.43
0.183	0.000408	0.694	1	0.296
0.620	0.00138	2.34	3.38	1

Turbidity / Visibility / Suspended Solids

NTU (Nephelometric turbidity unit): A measurement of the amount of light scattered / absorbed as it passes through water. This equates with underwater visibility, which depends on the amount of suspended solids (turbidity) in the water.

If the visibility is over 10', the turbidity is probably under 1 NTU (the drinking water standard). If the visibility is a hundred feet, the turbidity is probably below 0.1 NTU.[4]

Tank and Concrete Formulas[6]

r = radius
d = diameter
h = height
$\pi = 3.14$

Circumference of a circle = πd
Area of a circle = πr^2
Volume of a cylinder = $\pi r^2 h$
Surface area of a cylinder = $2(\pi r^2)+\pi dh$
Volume of a sphere = $4/3(\pi r^3)$
Surface area of a sphere = $4\pi r^2$
Volume of a hemisphere = $2/3(\pi r^3)$
Surface area of a hemisphere = $2\pi r^2$

Appendix B: Tank Loads and Structural Considerations

This section gets a bit technical. If this puts you off, not to worry—just skip it (if you're not building a tank or building one under 1,000 gal/3.8 m³); or skim it, and glean what you do from it (if you're building a tank up to 30,000 gal/ 113 m³.) If you're building a bigger tank, hire an engineer and an experienced contractor.

Forces on Tanks

A tank has to resist a variety of forces structurally:

- **Water pressure**—Acts at right angles to every surface—pushing down on the floor, out on the walls, *up* on the roof (if the roof space is filled with water). Water pressure is directly proportional to depth alone. If you are building your own storage with a water column deeper than 6', be careful. For water deeper than 8', you should have an engineer involved with the design. (See also Hoop Stress, next page.)

FIGURE 32: PRESSURE DEPENDS ON WATER DEPTH ALONE

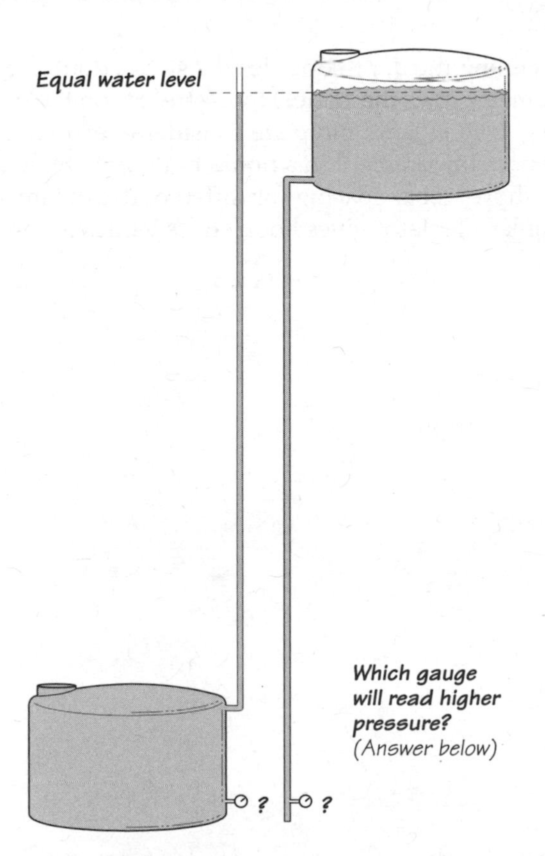

- **Point loads**—These include people walking on the roof and floor, rocks poking into the underside of the floor, cars running into the side, kids with ice picks, yahoos using your tank for target practice…
- **Earthquake loads**—These can topple the tank onto its side, slide it sideways off a footing, knock over water towers, or simply shake the bejeebers out of the tank until the sides split open.

Fiberglass tank split open by an earthquake on the big island of Hawaii.

WWW.PACIFICGUNITE.COM

- **Wind loads**—These can blow over an empty tank or (in the case of very high winds) spear it with flying debris or crush it with falling trees.
- **Gravity loads**—If the tank is supported unevenly, the floor can crack or the sides split under the strain. This can happen if the earth under the tank settles unevenly, is washed out from under the tank by water, turns to squishy muck, or heaves with frost or big roots.
- **Ice loads**—If thick ice forms on the water surface inside a tank, and then the water level falls, ladders and interior pipes can be ripped right out of the tank walls. If pipes, fittings, or the whole tank freeze solid, they can split open.
- **Soil loads**—Buried tanks can be subjected to intense inward and / or upward pressures (see Buried Storage, p. 31).

A flexible floor, stiff floor, cylindrical walls, and domed roof all work totally differently structurally:

- **All a flexible floor has to do is not puncture or tear**—The earth (or gravel on top of earth) supports a flexible floor. Providing it is flexible enough to conform to whatever degree of unevenness is present in the supporting ground, the force on it is minimal. It is not being stretched, bent, or sheared, just gently compressed between the water and earth. How gently? The floor of an 8' deep *(2.4 m)* tank is pressed down just 4 psi *(28 kPa)*, and the earth presses back an equal amount. Saran wrap could probably resist this force.
- **A stiff floor**—On the other hand, can develop tremendous bending forces if it is supported unevenly. If a portion of the floor of a big tank is cantilevered out over wet, squishy soil which is doing nothing to support the tank, it is up to the floor to resist many tons of water that is trying to crack the concrete slab. Ironically, an 8" thick *(20 cm)*, steel-reinforced concrete slab might fail in this circumstance, while a thin, flexible pond membrane probably would stretch to conform to the new shape and be fine.

If you make a stiff floor, you've got to also make it really strong. One way to make the floor stronger is to give it a conical or dish shape (see shape discussion, below, and the sidebar Tempting Floor Shape Innovations, p. 112).

- **Cylindrical walls are placed in tension (pulling) by water pressure**—Walls of a bought tank will be fine, as the manufacturer engineers them and your installation won't change the loading (unless the overflow plugs, causing the tank to be pressurized like the one that is about to explode in Figure 32). There is more on designing your own walls under Hoop Stress, below.
- **A domed roof takes advantage of the fact that materials are much stronger in compression (squeezing) than in bending**—People walking on a flat roof stress it through bending, while a domed roof supports people walking on it mostly in compression. The domed shape uses the strength of the material to best advantage, as most materials are much stronger in compression than bending. A conical roof (or floor) resists loads in a combination of bending and compression.

Hoop Stress

"Hoop stress" is the term for the loading on a cylinder which is being pushed out evenly in all directions from the inside. This push tries to stretch the cylinder walls in tension. This is the dominant way tank walls are loaded, by water pressure from the inside. The "hoop stress" on tank walls under pressure is proportional to the depth of the water and the diameter of the tank.

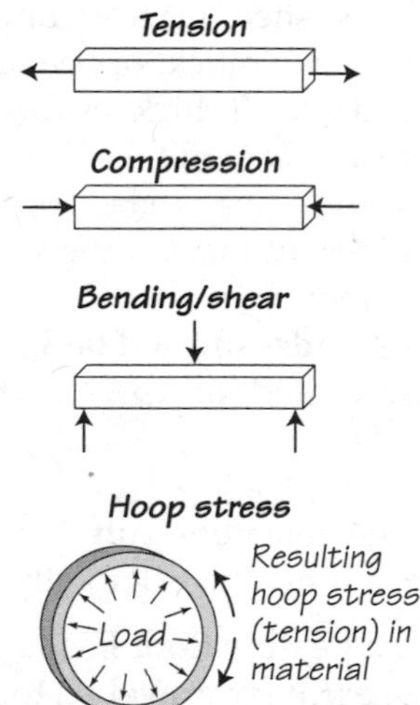

$$\sigma_h = \frac{pr}{t}$$

Where:
σ_h is the hoop stress
p is the water pressure
r is the tank radius
t is the wall thickness

If you are building a tank, you must consider the wall forces, and size the tension members of the lower wall to withstand the applied forces with a safety factor of at least two. If the walls are free to move at the bottom, they will be loaded in pure tension by outward water pressure (that is, the water will be trying to stretch the wall apart). Plastic and steel do very well resisting loads in pure tension. In fact, plastic, which is not an especially strong material, can withstand thousands of pounds of tension in a water tank.

In a redwood tank, the hoop stress is taken entirely by the hoops. The hoop vertical spacing is adjusted so all the hoops are equally loaded (see photo, p. 45). Because the walls in a redwood tank are free to move away from the floor slightly when the hoops are stressed, there is no shear between the floor and walls, and the force is just as in this ideal equation.

If the walls are rigidly attached to the floor, the loading at the bottom is a combination of tension, bending, and shear (the latter concentrated at the wall-to-floor joint). This requires extra reinforcement in the case of rigid materials such as steel or ferrocement. (Plastic just stretches a tiny bit extra until the shear is negligible and the material is mostly in tension.)

In a ferrocement tank, the walls can't shift away from the floor. They bend, putting shear stress in the material but reducing hoop stress. Also, lath and hardware cloth prevent even fine cracks in the plaster, so the plaster actually can (and does) carry quite a bit of the tension load. Because of these factors, the rebar spacing does not have to be as close as if it were taking 100% of the tension load.

There are three strategies to strengthen the base of the walls and the wall-to-floor joint in a ferrocement tank:

- **Increase the rebar spacing and size near the bottom of the wall.**
- **Increase the thickness of the wall near the bottom**—Using fatter rebar automatically increases the thickness of the plaster inside the armature, and the plasterers can be instructed to make the coverage thicker near the bottom.
- **Fatten the joint between the floor and wall inside and out**—With a fillet (radiused infill of the corner) of plaster.

Doubling the depth doubles the stress; doubling the diameter does the same. Hoop stresses on tanks of 10,000 gal *(38 m^3)* and more are considerable; do not skimp on reinforcement in a home-built tank. Figure 43, p. 114, shows rebar spacing for different depth tanks. Our Tank Calculator gives hoop stress values for any size tank.[6]

The Effect of Size and Shape

Size matters—The forces on big tanks are tremendously greater. A 1,000 gal *(3.8 m³)* tank can be built totally seat-of-the-pants by anyone. Medium sized tanks (10,000 gal/*38 m³*) require some head scratching. Large tanks (over 30,000 gal/*110 m³*) should be professionally engineered.

Not only are the forces greater, the consequences of failure are greater, too. I was asked to design a 100,000 gal (*380 m³*) tank in a steep, narrow valley in highland central Mexico. The tank would be located in an active earthquake zone a short distance from Popocatepetl, an active volcano. If it were to fail catastrophically, the resulting tidal wave would obliterate at least the nearest house, and possibly kill several people. Fortunately, I was able to instead help the community conserve enough water that the tank was not necessary.

Smaller things act much stronger—Because the loads on bigger things are proportionally bigger. Thus, a small diameter polyethylene pipe can contain a column of water a hundred feet deep *(30 m)*, while a big polyethylene tank can't be much taller than a person before the pressure bursts it.

Shape matters—The shape determines how the material will resist the applied force and thus how easy it will be to resist a given load. For example, a snow load bends a flat roof but puts the material in a domed roof under compression. A membrane roof would stretch concave, under tension. Here are some general guidelines for thinking about how shape affects structural integrity:

- **Compound curves are inherently strongest**—As in a dome, sphere, or urn.
- **Simple curves are strong**—As in cylindrical tank walls.
- **Folded planes**—Stronger than flat planes, as in a cone, or a peaked roof compared to a flat roof, or corrugated compared to flat sheet steel.
- **Triangles are stronger than straight pieces**—As in a roof truss compared to a beam.
- **Thicker, taller pieces**—Stronger than thin or flat pieces in bending. A load on a wide, flat piece of wood will bend it much more than it would bend the same piece stood on edge.

ORIENTATION CHANGES STRENGTH

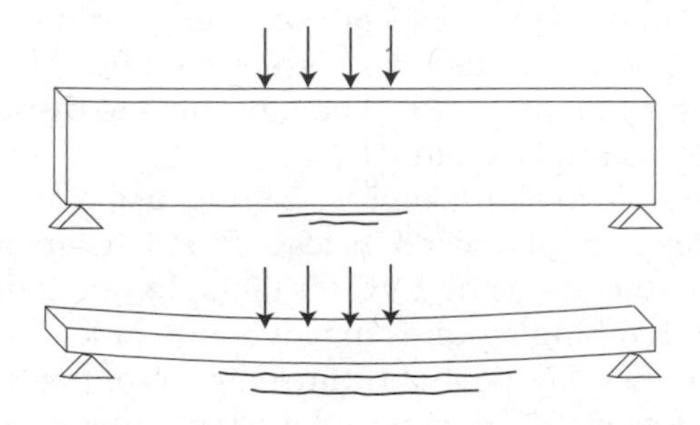

- **Uniformly stressed structures are stronger**—A wall that tapers toward the top and flares toward the bottom is stressed evenly, as there is more material where there is more stress. It would thus be harder to snap the wall off its footing than if it had uniform thickness throughout, in which case all the stress concentrates at the bottom.
- **Shorter spans are much stronger**—For example, a flat roof that is supported by a central pillar is four times stronger than one that spans the whole diameter of the tank without interior support.
- **Shapes that enclose more volume with the same amount of material are stronger, up to a point**—For example, a tube is stronger than a solid rod, and a bigger, thinner-walled tube is stronger than a smaller, thicker-walled tube. This holds true until the walls are so thin they buckle. *Note: This is true for compression, bending and twisting, but not tension. For tension, the shape doesn't matter, only the amount of material.*

Jar-style tank made of finger-thin, unreinforced sand/cement stucco. Compound curves and small size allow this tank to resist water loads and even careful transport in a truck. The bottom has the dish shape shown on p. 112. This one is 80 gal (300 L). Unreinforced jars can be up to 270 gal (1,000 L).

Appendix C: More About Plastics

This is supplemental information on the health and environmental issues with plastics used for water containers, storage, or plumbing. The more basic information and the details on the most commonly used materials can be found under Tank Materials/Plastic, p. 45, and in Table 7: Characteristics of Different Tank Materials, p. 40. If you really want to get into it, you can download the research notes behind this section.[6] These consist of pages and pages of abstracts of studies and web links on leaching from all materials (not just plastics), permeation, disinfection byproducts, and bacterial regrowth.

This section is organized roughly from most to least recommended material. The name is followed by the plastic's abbreviation and the recycling code number, common uses, recommended uses, and recyclability.

Biologically Based Polymers

Hopefully in the future, there will be water storage and plumbing components made of bio-based polymers, but I'm unaware of any available now.

High Density Polyethylene (HDPE #2)

HDPE is the preferred plastic for water tanks (for which it is the most commonly used material) and water lines, for which it is rare. It is relatively innocuous in its manufacturing, use, and disposal, at least compared to other plastics. (See Tank Materials/HDPE, p. 46, for more info.)

Low-Density Polyethylene (LDPE #4)

Used in food storage bags and some "soft" bottles. From a health and ecological standpoint, it is the same as HDPE (above).

Polypropylene (PP #5)

Used in rigid containers, including some baby bottles, and some cups and bowls. Comparable to HDPE in health and environmental pedigree.

Polyethylene Terephthalate (PETE/PET #1)

Used for most clear beverage bottles. The best combination of good taste, shatter-resistance, light weight, and acceptable environmental and health issues for small, portable water bottles of 1 gal or less.

Ethylene Propylene Diene Monomer (EPDM)

EPDM is commonly used for pond liners. It is a synthetic rubber resistant to heat, ozone, and UV light. It is able to stretch quite a bit without tearing. There is little data on leaching of EPDM, although it is generally considered to be pretty inert. It is considered a more environmentally friendly building material than PVC. (See Tank Materials/EPDM, p. 46, for more info.)

Polyamide Epoxy

There is evidence that epoxy coatings leach various toxic additives into water. The observed leaching decreases exponentially over time.

Polycarbonate/Lexan (PC #7-other)

Used in 5 gal water bottles, some baby bottles, some metal can linings, and popular Nalgene® backpacker's water bottles.

Valued for its high strength and the fact that it doesn't impart a taste to water. Unfortunately, polycarbonate can release its primary building block, bisphenol-A, a suspected hormone disrupter, into liquids and foods. This leaching worsens as the material ages and degrades.

Acrylonitrile Butiadene Styrene (ABS)

Not used for water tanks, or potable water plumbing. It is used for drain plumbing, and occasionally do-it-yourselfers will incorporate it in water supply systems. The manufacture of ABS generates hazardous materials, including carcinogens. ABS is difficult to recycle, and is considered only marginally better than PVC (see below) in terms of environment and health effects.

Cross-Linked Polyethylene (PEX)

Cross-linking makes PEX more durable than HDPE. There is concern that it leaches MTBE and benzene into water, and also that the pipes may prematurely decay and rupture, releasing flammable material that would allow a fire to quickly spread through the building. PEX should not be exposed to sunlight.

Fiberglass (Glass Fiber-Reinforced Polyester, GRP)

Fiberglass tanks are very strong, lightweight, and noncorrosive. Fiberglass is quite a bit stronger and more expensive than HDPE and generally considered to be higher quality. It is certainly superior to HDPE for underground tanks due to its high strength. Exceptionally nasty solvents are used in the resin used to make fiberglass. The literature is strangely silent on its health effects, but I'd be nervous about leaching of residual solvents.

Polyvinyl Chloride (PVC or V, #3)

Used in waterbed bladders and other flexible liners, some soft bottles, food cling wraps, and rigid pipe.

The manufacturing and installation impacts of PVC are so bad there is serious discussion of banning it. However, rigid PVC does seems to take a break from wreaking major environmental and health damage while it is in service. Rigid PVC in the shade does not seem to leach much into the water (but see discussion of issues with flexible PVC and PVC in sunlight, p. 39). When it is time to dispose of PVC, things get ugly again.

It is possible to eliminate PVC piping, using HDPE, copper, brass, or galvanized instead. Plumbers are universally adapted to using PVC for tank plumbing details, while HDPE tubing requires the invention or learning of new techniques. I hope that future editions of this book will show fewer PVC details and more of other types.

Appendix D: How to Make Ferrocement Tanks

Ferrocement tanks consist of an armature (framework) of steel reinforcing, covered with a sand-cement plaster. They offer near-complete flexibility in shape. They have a long life, are cost-competitive when contractor-built, and are owner-buildable in both industrialized and non-industrialized countries.

This section describes how to build ferrocement tanks with various techniques, to a variety of standards and sizes of 250–30,000 gal *(1–10 m^3)*. With the aid of an engineer you could adapt the plans up to a 100,000 gal *(380 m^3)* tank. For more on the advantages and characteristics of ferrocement tanks, see Tank Materials/ Ferrocement, p. 41.

The existing literature on ferrocement tanks is sparse, and each document is narrowly focused: one particular size of one design, or a variety of designs but all for the context of non-industrialized nations. The heavy-duty ferrocement construction technique described here—which is suitable for large tanks—has not been described in the literature before to my knowledge.

This appendix—practically a book in itself—is unique in that it describes the full range of ferrocement techniques in one place, and reconciles some enormously disparate opinions and techniques into a coherent formulary.

From procedures for ultra-light-duty tanks that use the absolute minimum of material, to tanks built successfully by native women with no construction experience, to detailed procedures for building large tanks to last a lifetime—you can glean the best approach for your context.

In the do-it-yourself, innovative spirit of ferrocement, this appendix gives you not only recipes but numerous variations and ideas for promising innovations, so that you can follow a recipe or concoct your own to suit:

- **Plans for Jumbo Thai Jar**—An 800 gal *(3 m^3)* light-duty cistern, which can be adapted to make containers of this shape of 250–800 gal *(1–3 m^3)*.
- **Description of ultra light-duty ferrocement**—For cisterns up to 3,000 gal *(11 m^3)* in size in the non-industrialized world.
- **Plans for light-duty ferrocement**—10,000 gal *(38 m^3)* cistern, adaptable for inexpensive, non-industrialized nation-style cisterns of 500–10,000 gal *(1.9–38 m^3)*.
- **Construction photos of medium-duty, urn-shaped, ferrocement cistern**—For a 3500 gal *(13 m^3)* cistern, which can be used in conjunction with the heavy-duty ferrocement construction plan to guide the construction of medium-duty construction cisterns of 500–15,000 gal *(1.9–57 m^3)*. These also illustrate how ferrocement can be used to make creative shapes and details.
- **Detailed plans for heavy-duty ferrocement construction**—For a 30,000 gal *(110 m^3)* cistern, which can be adapted to tanks of 3,000–30,000 gal *(11–110 m^3)* capacity, and with the aid of an engineer, tanks up to 100,000 gal *(380 m^3)*.

I suggest you read this whole appendix before designing anything. You'll see the huge difference that perfection standard and context make (see p. 4), and the even bigger difference that tank size makes. This will help you to make the perfect design adaptations for your context.

The heavy-duty ferrocement technique, which is by far the most complicated, has the most detailed instructions and drawings. Much can be learned from that section that will apply to simpler projects. For example, there is a complete tool list there. (All the tools on that list which don't have an asterisk are needed for the other methods as well as heavy-duty ferrocement.)

Don't Get in Over Your Head

Ferrocement requires a certain amount of manual skill, and any big construction project requires a certain amount of organization and management skill. Structural sensibility is a great help, too. (Reading Appendix B on structural considerations is highly recommended, especially for big tanks. Difficulties go up dramatically with the size of the tank.)

The instructions we offer here are fairly detailed, and will give you an enormous boost over figuring this stuff out on your own, or from any other source we know of. However, in a project of this nature, there is still going to be plenty of stuff you're just going to have to figure out. If you don't feel confident about taking on a tank-building project after reading this section, I suggest you consider buying or salvaging a tank instead of attempting to build one. If you feel only modestly confident, I suggest you start with a modest-sized tank—say 1,000 gal *(3.8 m^3)*—and see how that goes before sinking tens of thousands of dollars into the construction of a huge tank.

Design Innovation

Ferrocement construction is a relatively unexplored field, open for innovation and improvement. As I synthesized a wide variety of ferrocement techniques to create this appendix, cross-pollination between different techniques occurred. Some sources had, for example, brilliant structural analyses; others had ingenious time-saving construction techniques. I've exercised my best design discretion in taking insights from one source and applying them to other techniques. This is how I've handled it:

- **Unequivocal improvements with a low likelihood of significant implementation difficulties**—Have been put into the main narrative and drawings. In actual practice, there may be some minor differences. (If you think you've run into one, I hope you'll let us know.)
- **Promising but more experimental improvements**—That may require more than the usual level of head scratching to build successfully are given in the form of sidebars and separate drawings.

These prospective improvements have mostly been done in some form, but often not at the same scale or together in the same tank. For example, people have made buried half-sphere tanks in Africa, and people have made tanks with hemispherical roofs in California, but to my knowledge no one has put the two halves together to make a totally spherical ferrocement tank. Would some major unanticipated problem be encountered making a sphere? Seems unlikely, but there certainly could be.

You'll have to be the ultimate judge of the advisability of these innovations for your project.

I see this as a two-way forum for innovation. I hope you will take good photos of your project and let us know what worked and didn't, and what new ideas you come up with. We'll pay you for anything we broadcast through subsequent editions of this book or our website. (You can also check oasisdesign.net/water/storage for updates to this section.)

Ultra-Light Ferrocement over a Form: Jumbo Thai Jar Plans

These plans from the Intermediate Technology Development Group[18] are for the construction of an 800 gal (3 m^3) tank. It can be scaled to build tank sizes from 250 to 800 gal (1–3 m^3). The shape is extremely efficient structurally and for use of materials, and it's beautiful, too. This construction system has been highly successful in non-industrialized nations, especially Thailand, where there are millions of the 250 gal (1 m^3) version of this tank. It can be done with minimal masonry skills. For sizes 250 gal (1 m^3) and smaller, you can omit the steel reinforcement.

A technique common in non-industrialized nations is to use a form for backing. A form enables minimally skilled masons to do the work, enables the mix to go on more thickly and evenly, and helps slow curing (one side is kept from losing water by the form, and you can wrap the other with plastic). There are many form systems, used for tanks of up to a few thousand gallons *(10 m^3)*. One form system involves pouring concrete between two sets of steel shutters (see photo, p. 44). Another involves stretching chicken wire and thick wire over a corrugated, galvanized, one-sided form. The drawback of using a form is that you've got to make it, and the economics are better if it can be reused for many tanks.

My favorite form system is the Thai Jar, which incorporates a shape with strength-enhancing compound curves, and has a narrower diameter near the bottom to reduce hoop stress. The form is so simple that it could make sense even if you're only making one tank. In sizes of 250 gal *(1 m^3)* and smaller, it can be made without steel reinforcement. In sizes of 55 gal *(200 L)* and smaller, it can be made in the same shape out of fired clay.

Sew the Mold

The mold (form) is made of 15 m of canvas 1.2 m wide, consisting of five side panels and one floor panel sewn together with strong thread and overlaps of 10 cm on all joints (see Figure 33).

To make a mold, start by drawing a side panel in full size on the canvas and cut it accordingly. Use this side panel to mark the other four side panels and cut them. Finally, cut the floor panel, and then sew all six panels together.

Make space for a string to be pulled and tied for the manhole at the top of the mold.

Foundation

The jar edge should be situated 90 cm from the wall of the house (if it is to be used for rainwater harvesting). The radius of the foundation is 75 cm. Draw the circumference of the foundation using a string tied to a peg at the center point of the jar.

Dig out soil within the circle until firm soil is reached, or the height of the eave of the roof is 220 cm. Level the excavation (or make it dome-shaped—see p. 93, 112).

Fill the excavation with 10 cm of concrete 1:3:4 (cement: sand: gravel); make it level and compact it well.

FIGURE 33: 3 M³ JAR MOLD

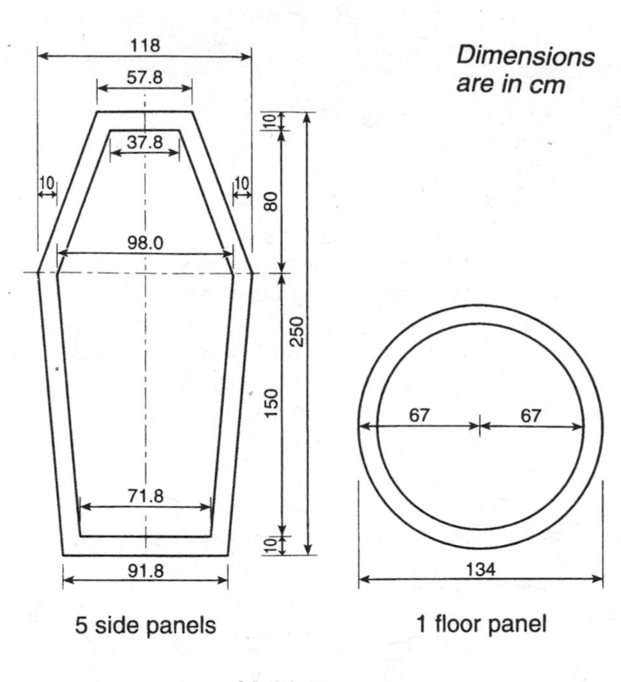

Mold

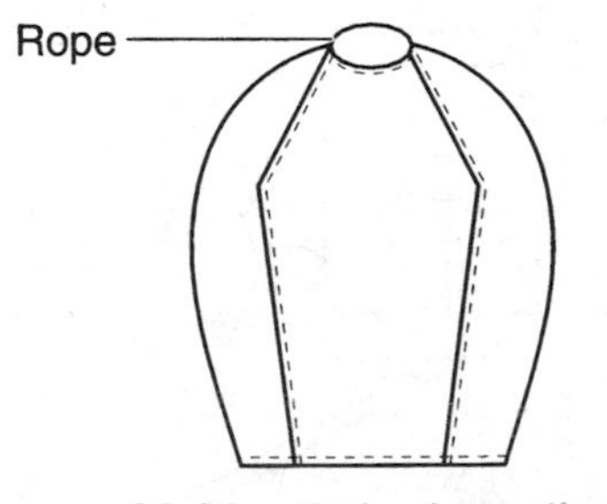

Mold stitched together

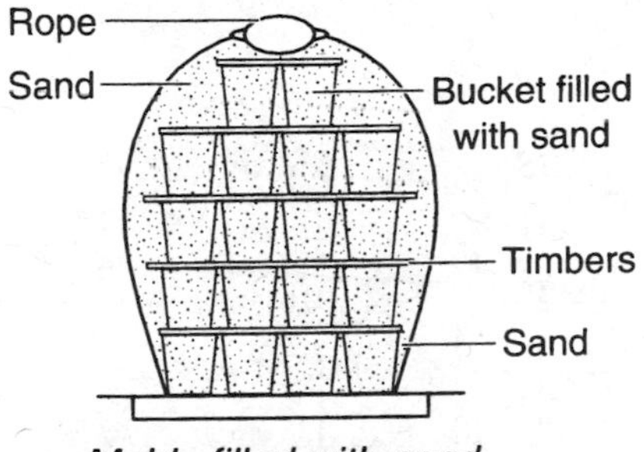

Mold filled with sand

TABLE 10: BILL OF MATERIALS FOR 3 M³ FERROCEMENT JAR

Item	Specification	Units	Qty
Materials			
Cement	50 kg *(110 lb)*	Bags	6
Lime	25 kg *(55 lb)*	Bags	1
Sand	Coarse and clean	Tons	3
Crushed stone	10–20 mm *(.39"-.79")*	Tons	1
Rubble stones	100–500 mm	Tons	1
Bricks/blocks	Variable	Number	50
Water	200 liters *(53 gal)*	Oil drums	3
PVC pipe	50 mm *(1.97")*	Meters *(Feet)*	3
Galvanized pipe	38 mm *(1-½")*	Meters *(Feet)*	0.5
Galvanized pipe	18 mm *(¾")*	Meters *(Feet)*	0.9
Tap, elbow, nipple, and socket	18 mm *(¾")*	Unit	1
Galvanized wire	3 mm *(1.18")*	Kg	5
Chicken mesh	25 mm *(.98")*, 0.9 mm *(.04")*	Meters *(Feet)*	18
Mosquito mesh	Plastic	Meters *(Feet)*	0.5
Hardware cloth	Galvanized 5 mm *(.20")*	Meters *(Feet)*	1
Canvas reusable for 10 jars	1.2 m *(3.94')* wide sewn into a mold	Meters *(Feet)*	15
Labor	Skilled masons	Working days	1×5
	Laborers	Working days	1×5
Total cost $150 *(2005)*	Labor and materials		

The outlet pipe is made of 90 cm of 18 mm galvanized iron pipe onto which an elbow and a nipple are screwed to the inner end, and a socket and a tap to the outer end. Place the pipe upon the foundation.

Floor Reinforcement

Cut eight lengths of 7 m-long 3 mm galvanized wire. Bend the wire ends to avoid injury. Mark the middle of each wire. Tie the eight wires together at the marks as spokes in a wheel. Make a ring of 3 mm galvanized iron wire, 116 cm in diameter, and tie it on the spokes.

Tie two 136 cm lengths of chicken wire with overlaps of 10 cm to the ring of wire. Place the wires and mesh on the foundation.

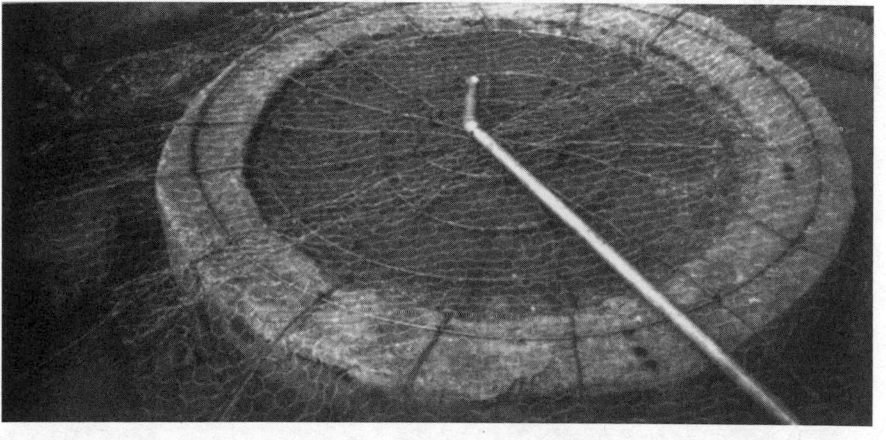

Fill the Mold

The mold is then placed on the foundation and stuffed with light, dry materials, e.g., sawdust, hay, or dung. Sand may be used. But since 3 m³ of sand weighs about 5 tons, a mold made of canvas will burst unless most of the sand is kept in buckets, stacked on layers of timbers separating the rows of buckets (see Figure 33).

Wall Reinforcement

Chicken wire is wrapped tightly around the stuffed mold while the chicken wire lying under the mold is bent up against the side of the mold.

The 16 wires sticking out from under the mold are now tied on to a ring of wire at the top of the mold and spaced equally.

The end of a roll of 3 mm wire is tied on to the foundation and wrapped tightly around the mold as a spiral spaced 20 cm from the top of the mold.

External Plaster

Plaster 1:3 (cement: sand) is smeared onto the mold in a thin layer. After a couple of hours, more plaster is applied to the mold until the plaster is 2 cm thick.

While the plaster cures for three days, the tap station is built.

Internal Plaster and Finish

After three days, the mold and its contents are removed. The jar is cleaned before plaster 1:3 is applied to the internal side of the jar in two layers, each layer being 1 cm thick. The floor is made of 5 cm-thick plaster of 1:3. On the same day, cement and water are mixed with Nil (a mixture of cement and water the consistency of porridge) and, with a steel trowel, pressed into the moist plaster for waterproofing.

Place two concentric rings of plain sheet metal 10 cm high and 60 cm diameter on top of the jar. Fill the space with 1:3 plaster to form a manhole and a lip. Place a pipe for an overflow through the lid. Cover the manhole with mesh to prevent insects and debris from entering the jar.

FIGURE 34: FERROCEMENT 3 M³ THAI JAR SECTION AND PLAN
DIMENSIONS IN CM

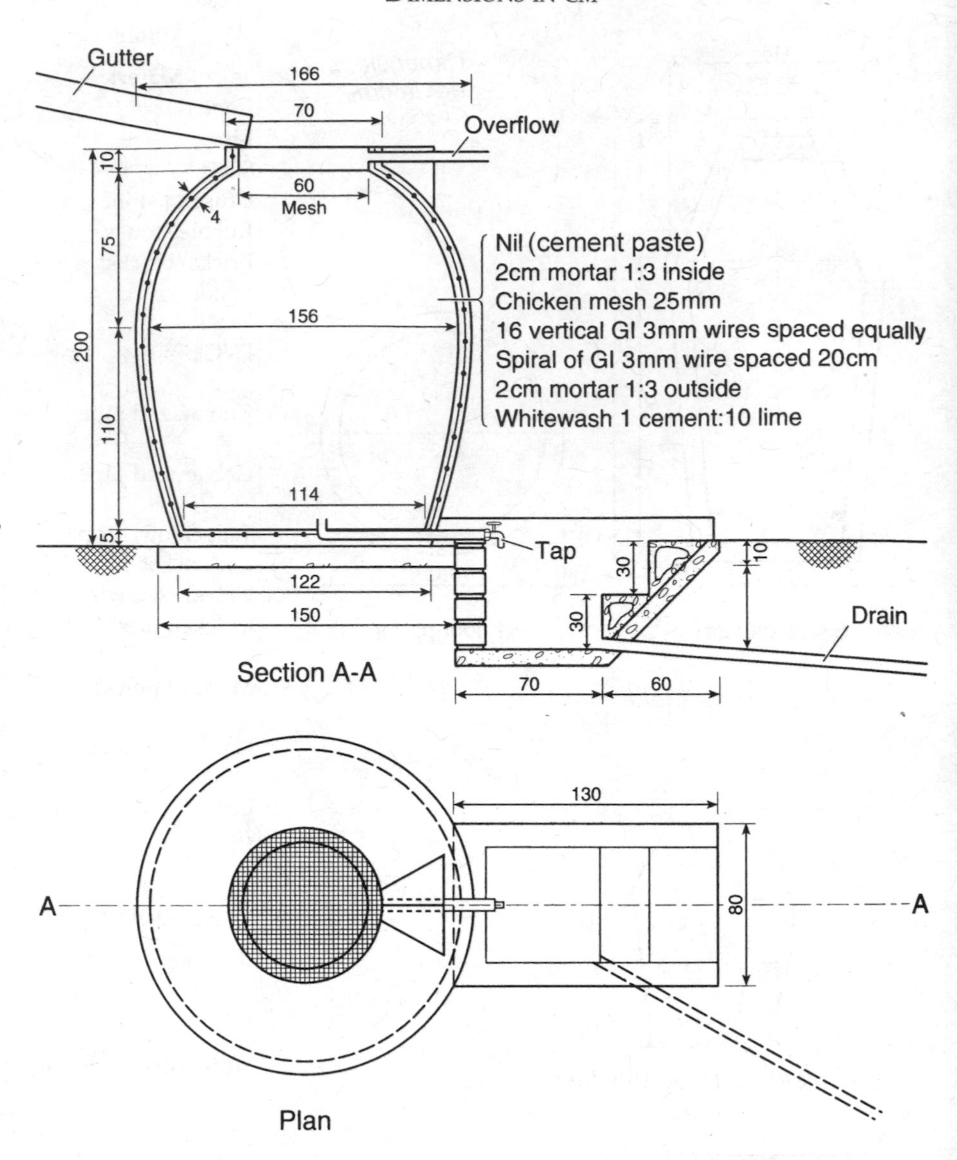

Metric	US
3 mm	0.12″
18 mm	0.71″
25 mm	0.98″
1 cm	0.39″
2 cm	0.79″
5 cm	1.97″
10 cm	3.94″
20 cm	7.87″
30 cm	11.81″
37.8 cm	14.88″
57.8 cm	22.76″
60 cm	23.62″
67 cm	26.38″
70 cm	27.56″
71.8 cm	28.27″
75 cm	29.53″
80 cm	31.5″
90 cm	35.43″
91.8 cm	36.14″

Metric	US
98 cm	38.58″
110 cm	43.31″
114 cm	44.88″
116 cm	45.67″
118 cm	46.46″
122 cm	48.03″
134 cm	52.76″
136 cm	53.54″
150 cm	59.06″
156 cm	61.42″
166 cm	65.35″
200 cm	78.74″
220 cm	86.61″
250 cm	98.43″
7 m	22.97′
15 m	49.21′
3 m³	105.94 ft³

FIGURE 35: 3 M^3 JAR CONSTRUCTION

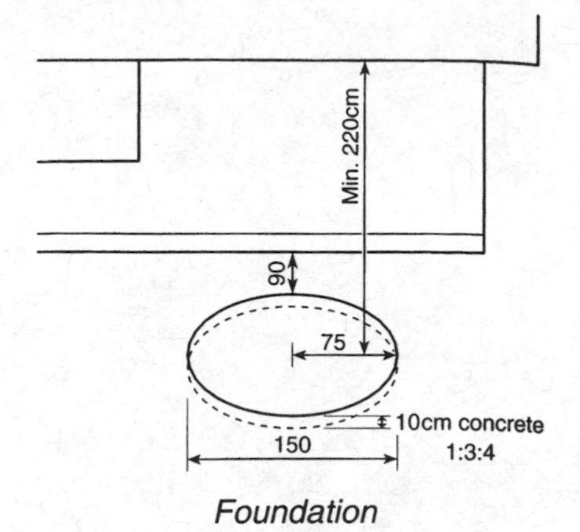

Foundation

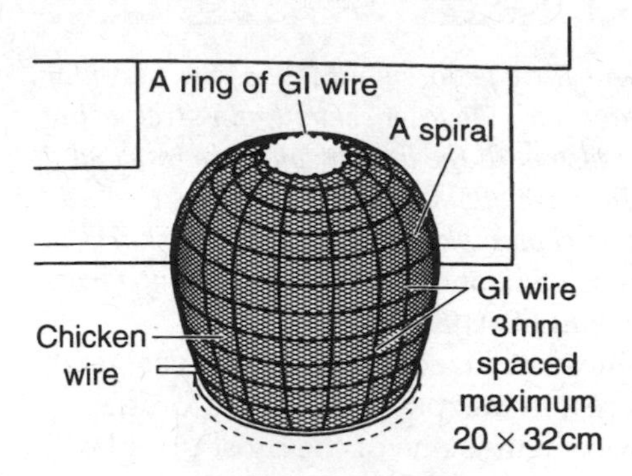

Reinforcement of wall

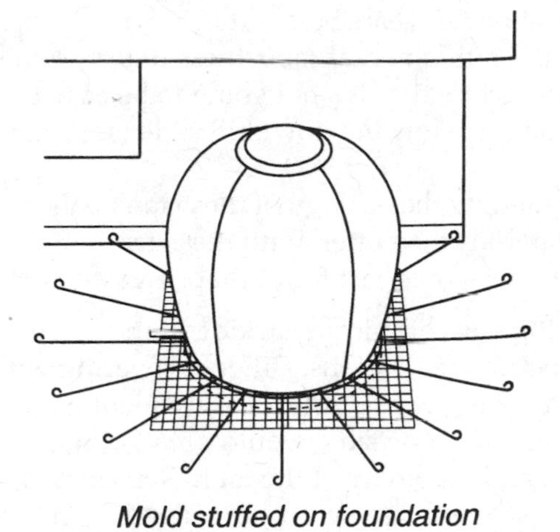

Reinforcement of floor

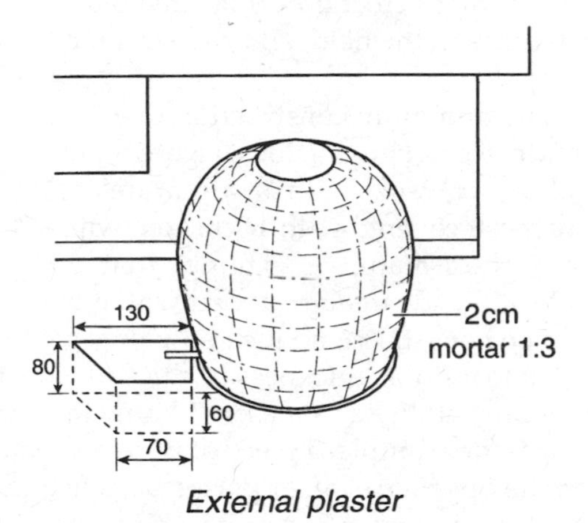

External plaster

Mold stuffed on foundation

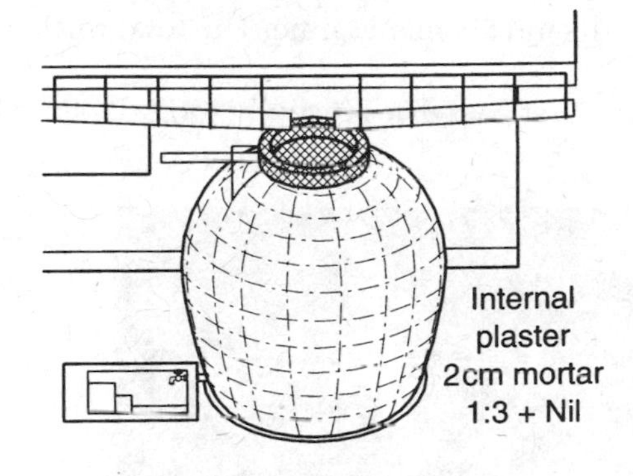

Internal plaster and finish

Metric	*US*
3 mm	*0.12″*
2 cm	*0.79″*
10 cm	*3.94″*
20 cm	*7.87″*
32 cm	*12.60″*
60 cm	*23.62″*
70 cm	*27.56″*
75 cm	*29.53″*
80 cm	*31.5″*
90 cm	*35.43″*
130 cm	*51.18″*
150 cm	*59.06″*

Ultra-Light-Duty Ferrocement Description

Ultra-light-duty ferrocement is the way to contain the most water with the least cement and steel, using no form, but also with the lowest safety factors and durability, in tanks of 500–2,000 gal (1.9–7.6 m^3). It's an appropriate choice where there are fairly skilled, patient masons and not enough money to buy a more adequate amount of material.

In non-industrialized nations there is a great deal of interest in inexpensive water storage tanks. Ferrocement is one of the most popular options. Some of these systems take the amount of reinforcement and materials right down to the absolute minimum. Even at this, they are holding up for decades in the field, and catastrophic failures are rare.

Simple, inexpensive owner-built ferrocement tank in an Indian village in Mexico, after first coat, and before second. Note how you can see the reinforcing pattern showing through the inadequate cement coverage.

The absolute minimum construction I've seen was in Mexico, where the locals were building 2,000 gal (7.6 m^3) tanks with an armature of welded wire mesh covered with 1" chicken wire—nothing more. The armature is so floppy that it has to be held still and circular-shaped with guy wires or strings. To be honest, I've no idea how they get the plaster to stick to a single thickness of chicken wire; this is well beyond my ability as a mason. I know they use some backing—for example, a piece of plywood held by someone on the opposite side, or plastic bags tied to the outside. This technique requires quite a bit of skill, or much of the plaster will end up on the ground. The mix is about 3:1:1 sand : cement : lime. The total thickness is ¾" to 1–¼".

After this first coat dries (usually too fast, since it is just barely hanging there in the breeze, and since most masons in non-industrialized nations aren't tuned into wetting masonry to slow the curing), another coat is added inside and outside.

Here a 20 year old, ultra-light-duty ferrocement tank gets a re-plastering in an effort to slow its disintegration, with mixed results.

Considering the overly fast curing, abundance of undesirable cold joints (where cured cement meets fresh cement), likely leak points, minimal reinforcement, and often minimal cement cover over the steel, it is surprising these tanks hold water. That hundreds of them do, and do so for about 20 years before they start to leak, is testimony to the fundamental robustness of ferrocement.

Once they start to leak, it's not going to be easy to repair them, only to slow the downhill slide (see photos at left).

I am intrigued by the ecology of these tanks, the very low use of materials they offer. With these few refinements, I think they could last for perhaps twice as long:

- **Proper curing**—Is chronically lacking in the masonry of non-industrialized nations. Diligent covering with rags and wrapping with plastic, and frequent wetting, especially on the sunny side, would greatly improve the strength and longevity of the tank. Reused pallet wrap (giant, strong Saran wrap, scavenged from the receiving dock of some warehouse) is the ultimate slow curing aid.
- **Thicker mortar**—The failure mode seems to be corroding of the chicken wire. Having at least 1 cm of mortar coverage over all steel would greatly extend the life of the tank. As a practical matter, this can probably only be achieved by doing one or more extra coats of plaster inside and out—the initial coats can only be so thick, as they aren't really attached to anything. These could be done rapidly with unskilled labor, using a soupy mix applied with a mason's brush.

❖ **Extra reinforcement at the bottom**—To contain higher pressure there and reduce the chance of leaks in this most inconvenient of locations. Use perhaps three turns of thick annealed wire, a second layer of welded wire mesh, or one or two hoops of rebar around the bottom 20% of the tank. This would almost double the pressure resistance for little added cost.

These refinements are incorporated in the Light-Duty Ferrocement Plans, below. The construction procedure is similar to that in the Heavy-Duty Ferrocement Plans described in great detail later, except that there are two layers of reinforcement instead of several.

If you are not a skilled mason and want to attempt one of these, use expanded metal lath instead of chicken wire, and you will have a fighting chance.

FIGURE 36: 46 M³ LIGHT-DUTY FERROCEMENT TANK SECTION AND PLAN (DIMENSIONS IN CM)

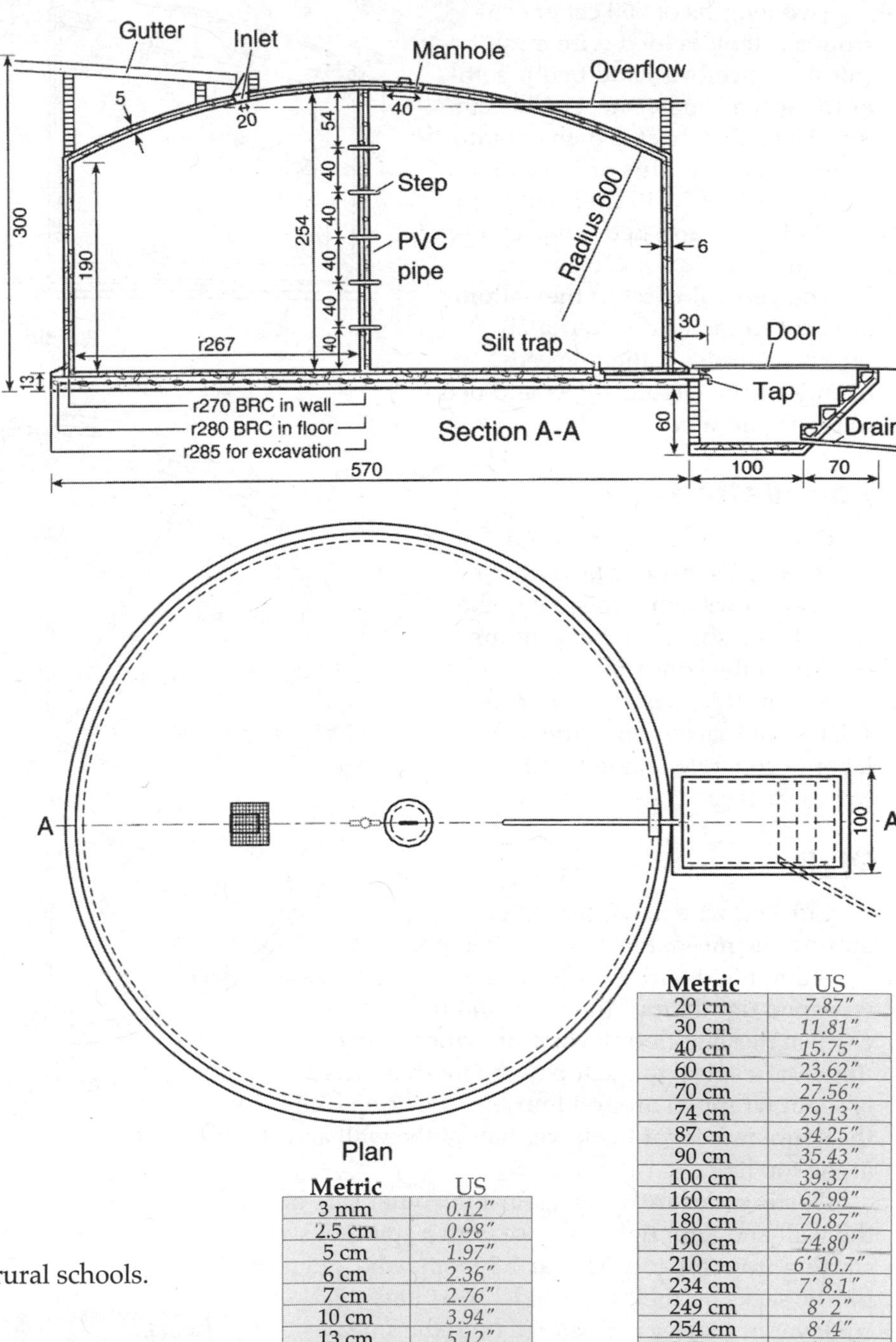

Metric	US
3 mm	0.12″
2.5 cm	0.98″
5 cm	1.97″
6 cm	2.36″
7 cm	2.76″
10 cm	3.94″
13 cm	5.12″
15 cm	5.90″

Metric	US
20 cm	7.87″
30 cm	11.81″
40 cm	15.75″
60 cm	23.62″
70 cm	27.56″
74 cm	29.13″
87 cm	34.25″
90 cm	35.43″
100 cm	39.37″
160 cm	62.99″
180 cm	70.87″
190 cm	74.80″
210 cm	6′ 10.7″
234 cm	7′ 8.1″
249 cm	8′ 2″
254 cm	8′ 4″
267 cm	8′ 9.1″

Light-Duty Ferrocement Plans

Plans from the Intermediate Technology Development Group[18] for somewhat more reinforced, but still lightweight tank of 12,000 gal (46 m³), the maximum size I'd make without rebar. They can be scaled to build tank sizes from 500–12,000 gal (2–46 m³). This system has lower safety factors and durability than medium- or heavy-duty construction, and is most suited to non-industrialized nations, where money for material is very tight. It requires fairly skilled masons.

Approximately 4,000 such tanks have been built in Kenya, to provide harvested rainwater for drinking water at rural schools.

Excavation

If the tank is to be used for rainwater harvesting, the edge of the excavation circle should be at the mid point of the gable roof of the house, 90 cm from the house wall and at least 300 cm below the eave of the roof. In any case, draw a circle from the midpoint of the tank, with a radius of 285 cm. The excavation should be at least 300 cm below the eave of the roof and at least 15 cm deep, or until firm soil is reached. Make the floor of the excavation level.

Welded Wire Mesh for Floor and Wall

Two lengths of 560 cm are cut from a roll of welded wire mesh (mesh, henceforth) and tied together to form a square sheet of 560 cm × 560 cm. The sheet is then cut into a circle with a radius of 280 cm.

A length of 1,740 cm is cut from a roll of mesh and tied into a cylinder with a radius of 270 cm.

The vertical wires at the bottom are bent to each side alternately.

The cylinder is then placed evenly on the circular sheet and tied to it with tie wire.

FIGURE 37: 46 M³ FERROCEMENT TANK ROOF

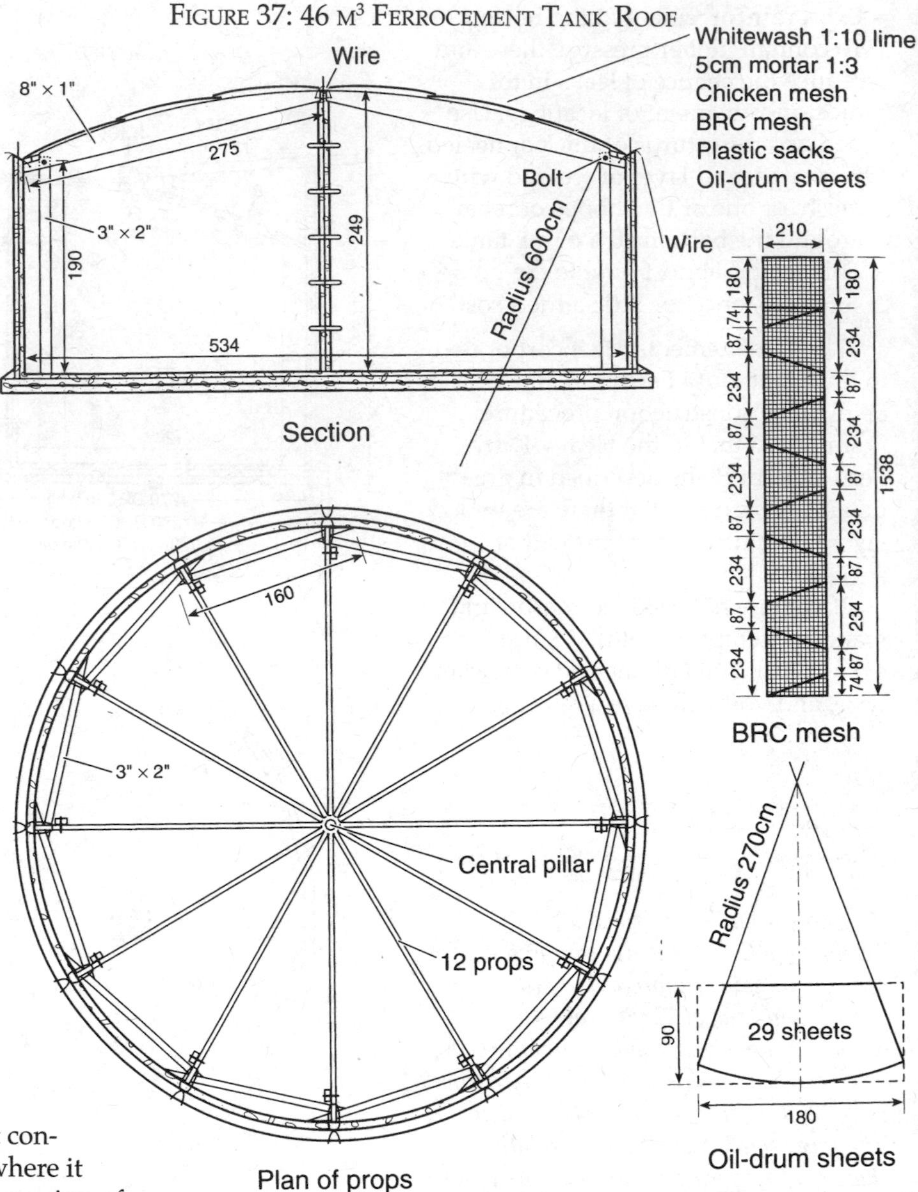

Foundation

Concrete 1:3:4 is mixed and placed in a 7 cm-thick layer in the excavation without moistening the soil. The mesh and outlet pipe are placed on the concrete.

A 6 cm-thick layer of concrete 1:3:4 is compacted on to the first layer of concrete and left with a rough surface.

Wall

Chicken wire is wrapped tightly around the mesh, twisted, and tied on.

A 3 mm galvanized iron wire is wrapped tightly four times around the chicken mesh at floor level, from where it continues as a spiral to the top of the mesh, where it is again wrapped around four times. The spacing of the spiral is 5 cm at the lower half of the wall and 10 cm at the top half.

Plastic sacks are hung against the outer side of the wall and kept tight in place with a spiral of string starting from the top. Mortar 1:3 is smeared against the plastic sacks on their inner side. Next day, a 2.5 cm layer of plaster is plastered onto the smear and floor and finished with nil.

The sacks are removed and the outer wall is plastered with 2.5 cm of plaster 1:3 sand: cement.

Metric	US
270 cm	*8′ 10.2″*
275 cm	*9′ 0.3″*
280 cm	*9′ 2.2″*
285 cm	*9′ 4.2″*
300 cm	*9′ 10.1″*
534 cm	*17′ 6.2″*

Metric	US
560 cm	*18′ 4.4″*
570 cm	*18′ 8.4″*
600 cm	*19′ 8.2″*
1538 cm	*50′ 5.5″*
1740 cm	*57′ 1.0″*

Dome

Erect the formwork and cover the plastic sacks with mesh. Bend the vertical mesh ends in the wall over the mesh in the dome. Lightly compact a 5 cm plaster 1:3 onto the dome while lifting the mesh into the middle of the plaster.

Use a washbasin as a form for the manhole.

Make 20 cm × 20 cm inlet holes.

FIGURE 38: 46 m^3 TANK CONSTRUCTION

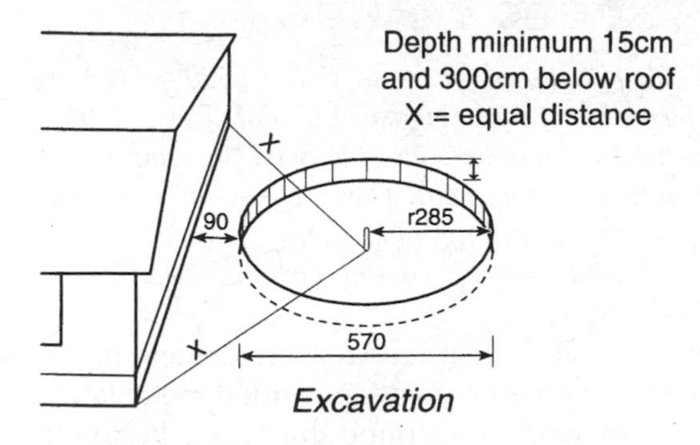

Excavation

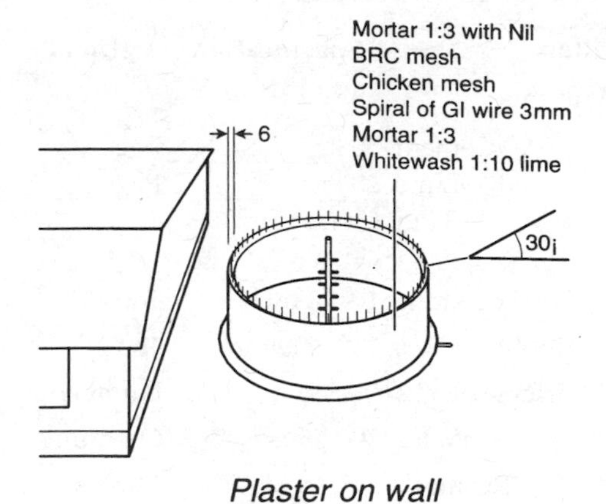

Plaster on wall

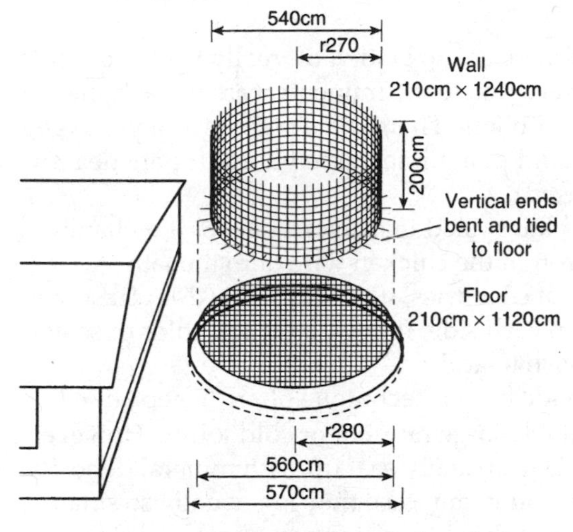

Mesh for floor and wall

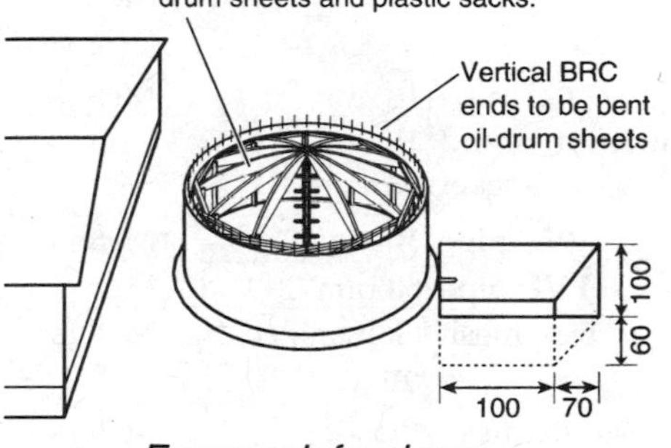

Formwork for dome

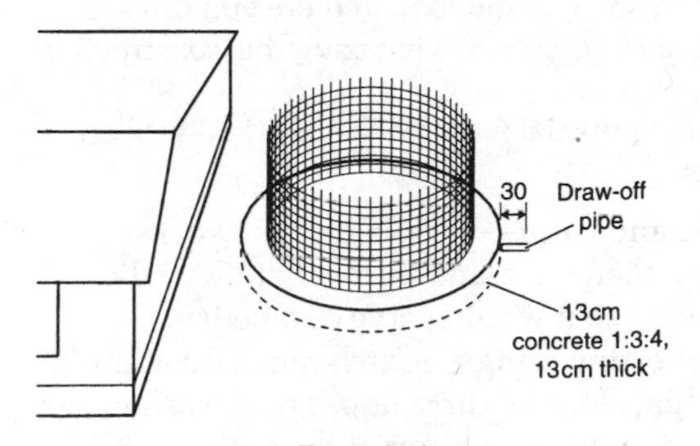

Foundation

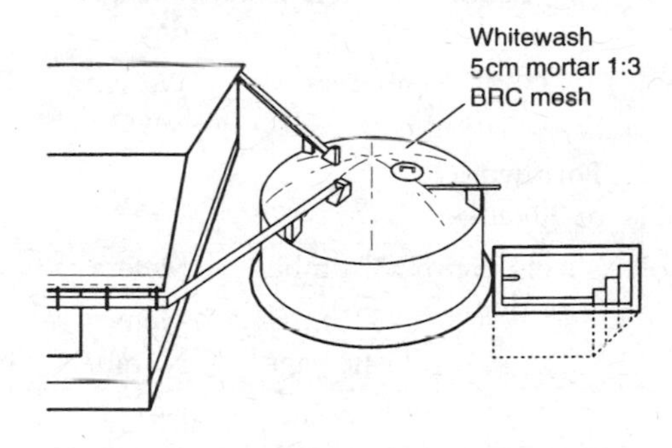

Dome, manhole, overflow and tap

Metric	US
6 cm	2.36″
13 cm	5.12″
15 cm	5.90″
20 cm	7.87″
30 cm	11.81″
60 cm	23.62″
70 cm	27.56″
90 cm	35.43″
100 cm	39.37″
200 cm	6′ 6.7″
210 cm	6′ 10.7″
270 cm	8′ 10.3″
280 cm	9′ 2.2″
285 cm	9′ 4.2″
300 cm	9′ 10.1″
540 cm	17′ 8.6″
560 cm	18′ 4.4″
570 cm	18′ 8.4″
1,120 cm	36′ 8.9″
1,240 cm	40′ 8.2″

Cover the finished dome with plastic sacks weighed down by sand or soil. Do not walk on the dome for seven days, after which the formwork can be removed.

Inlets, Overflow, and Tap

Build the inlets and install the overflow pipe over the tap-stand, which can be closed with a door. Seal the joint between dome and wall.

TABLE 11: BILL OF MATERIALS FOR A 46 M³ FERROCEMENT TANK

Item	Specification	Units	Qty
Materials			
Cement	50 kg	Bags	50
Lime	25 kg	Bags	2
Sand	Coarse and clean	Tons	10
Crushed stone	10–20 mm	Tons	4
Rubble stones	100–500 mm	Tons	1
Bricks/blocks	Variable	Number	50
Water	200 liters	Oil drums	35
BRC mesh	No. 65	Meters	33
Chicken mesh	25 mm, 0.9 mm	Meters	80
Twisted iron	12 mm *(½")*	Meters	3
G.I. wire	3 mm	Kg	25
G.I. pipe	37 mm *(1-½")*	Meters	1.8
G.I. pipe	18 mm *(¾")*	Meters	3.4
Tap, elbow, nipple, and socket	18 mm *(¾")*	Unit	1
PVC pipe	100 mm *(4")*	Meters	3
PVC pipe	50 mm *(2")*	Meters	3
Fine mesh	Galvanized 5 mm	Meters	1
Mosquito mesh	Plastic	Meters	0.5
Lockable door	Steel	0.9x1.5 m	1
Labor	Skilled masons	Working days	2x14
	Laborers	Working days	3x14
Formwork Reusable for 30 tanks			
Timber, bolts, sheets, etc. for dome	6"x1" timber	Meters	36
	2"x3" timber	Meters	46
	Plastic bags	Number	50
	Sisal twine	Kg	5
	Bolts 6x120 mm	Number	12
	Oil-drum sheets	Number	29
Manhole	Plastic basin	Number	1
Cost $1,200 *(2005)*	Materials and labor		

Medium-Duty Shaped Ferrocement Photos

Many people have asked for plans for our 3500 gal (13 m³) urn-shaped ferrocement tank (big cover photo). These photos show its construction, and how you can bend the armature to get unique shapes in ferrocement. The construction technique shown is applicable to building tanks from 1,000–15,000 gal (3.8–5.7 m³), in industrialized and non-industrialized countries.

The most accessible technique for ferrocement in the industrialized world is the use of expanded metal lath over a rebar framework. I've found this works best with chicken wire on the other side. The minimum reinforcement is a grid of 3/8" rebar, about a foot on center both ways.

Everything needs to be tied off really well and tight, which is a very time-consuming process when done with wire and pliers. There is a certain technique to getting it tight and doing it fast (see photos for an idea to get you started).

A somewhat skilled mason can plaster it by hand, pushing through the chicken wire onto the lath. After this structural coat dries, it generally needs at least one other coat on each side, followed by any color or sealer coats you wish to add.

Tanks made by this technique often "weep" due to the unavoidably large number of cold joints. However, these small leaks usually seal up with mineral deposits before long, and in any case they are usually so small they don't even drip.

A similar technique is often used to make artificial rocks and pools in zoos. The inside of the "rocks" is often left unfinished; if you look in there you can see the lath and rebar. Again, it is amazing this construction lasts, but it does.

Some of the unusual features that worked well on this tank are:

- **The shape and color**—The neighbors like it so much more than we dared think, that I wish I'd made it taller, so it would serve as a hedge.
- **Water-harvesting wings**—Catch rain, and allow about 300 gal *(1 m³)* of dirty house roof wash water to be stored *separately*, on top of the tank.
- **Many inlets and outlets made from PVC**—With sawn-in crisscrosses, none of which leak.
- **Hidden inlet pipe**—Passes through the tank floor and roof.
- **Overflow**—Goes in a wide, thin waterfall across several feet of wing (looks very cool).
- **Sloped floor and drain sump**—Make cleaning very easy.
- **Reduced visual mass**—Due to partial burial.

1. The site, before digging.

2. Grading the excavation with slope to the drain, using a water level.

3. Bending and tying rebar for the floor (note—wall in background is ferrocement, too).

4. Concrete filling "X" of trenches makes grade beams to reinforce floor.

5. Pouring the floor.

6. Leveling the floor to 2% slope and finishing.

7. Bending the rebar verticals.

8. Circle drawn on driveway as guide for shape of horizontal rings of rebar.

9. Placing the first rings and then verticals.

10. Bending and tying off the roof rebar.

11. Tying off lath (inside) and chicken wire (outside); first make a hook on the end of the wire, push it through...

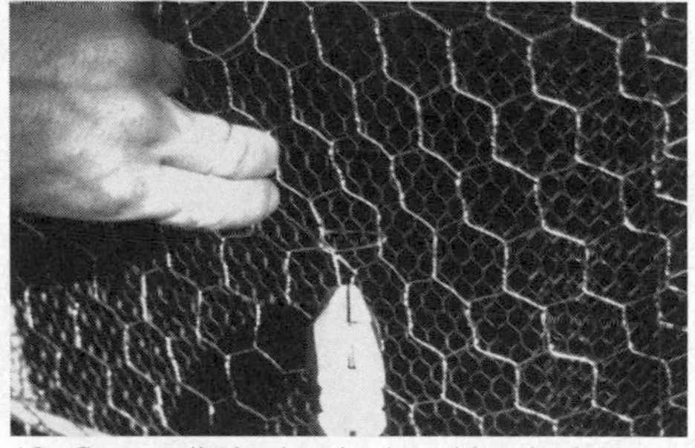

12. Then pull the hooked end back through, grab it with linesman's pliers, twist, then cut the tie off your spool of wire.

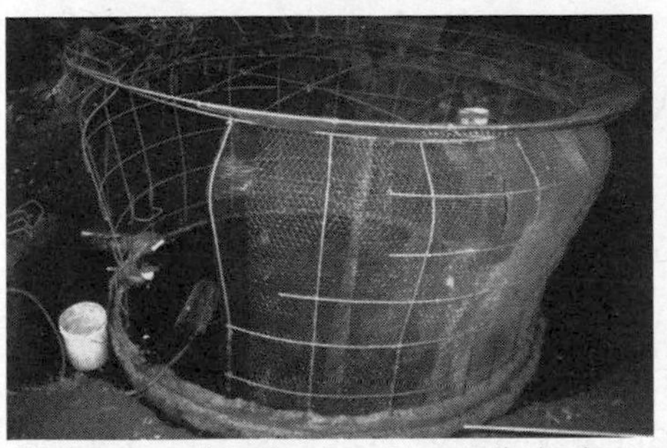

13. Leave a door in the side until the last minute; save lots of awkward climbing.

14. Check that every inch is tied off tight.

15. Plaster the built-in ladder first and when it dries you'll be able to get in and out easier after sealing off the temporary door.

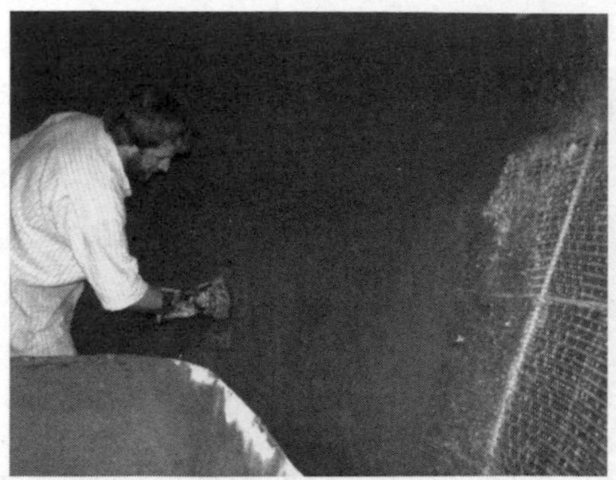

16. Plaster inside.

17. Plaster outside.

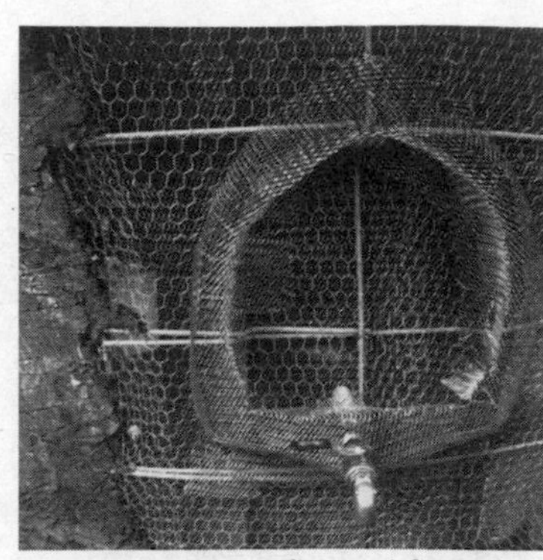

18. Altar detail over tap, formed from lath.

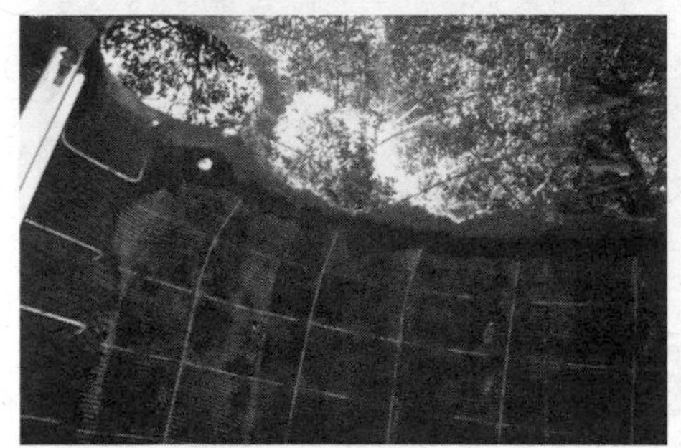

19. Lath roof looks cool.

20. White Thoroughseal to seal inside. Pipes are concealed inlets on way to roof.

21. Check out your plaster colors carefully, with big swatches... we measured to the gram on a postal scale, then multiplied for big batches.

22. Apply color coat (see color photo on front cover).

Ideas for Doing This More Easily than We Did

The way we made this tank—plastered all by hand, and with lots of details—caused it to be very labor-intensive. The curves and sculptural details are inherently more work, but other extra time was unnecessary. If I were to do it again, I'd make the following changes, most of which are described in the Heavy-Duty Ferrocement plans on the next page.

- **Join lath and chicken wire with hog rings**—Instead of wire.
- **Mix the plaster in a machine and apply it by pump**—The armature might need a layer or two of welded wire mesh to make this practical; it was on the floppy side and if you shook it or put plaster on too thick, it would slump off.
- **Make a smaller construction access passage**—And seal it up before plastering.
- **Use regular rebar**—Instead of epoxy-coated rebar, which is great stuff but probably overkill if the cement coverage is at least ½" *(1.25 cm)* and there is lath or chicken wire on both sides.

Heavy-Duty Ferrocement Plans

These are detailed plans for constructing a 30,000 gal (110 m^3) tank, which can be scaled from 3,000 to (with the help of n engineer) 100,000 gal (11–380 m^3). This technique has the ighest safety factors and longevity. It is most suited to an ıdustrialized nation context and tanks of greater than 30,000 al in non-industrialized countries.

Since all the masonry can be hired out, in theory ıis kind of tank could be built even if you don't have ny masonry experience at all. I'd say the gamble in ıaterials for a large tank seems too big to me to attempt ithout extensive prior construction experience.

People who make ferrocement tanks for a living—es-ecially large ones—quickly tired of all the tedious labor nd have come up with some innovations which con-erve a huge amount of time and improve the results. he main innovations are:

A thicker, stronger armature—That works for tanks up to 100,000 gal and can support wet plaster applied from the top down.
Numerous time-saving details—Such as the use of hog ring pliers (manual and pneumatic) in place of pliers and wire.
Plastering the whole thing in one day—To make a totally cold-joint-free tank, using a plaster mixer and a pump (or a large crew doing it by hand).

The extra materials cost of heavy-duty ferrocement is onsiderable, but not only does this method save a great eal of time; it makes a stronger, more waterproof tank y eliminating cold joints. If you can afford the extra ıaterials in the armature and hiring a crew to plaster it, ıis is by far the preferred way to make a ferrocement ınk of 3,000–100,000 gal capacity.

We're going to describe the tools and materials eeded, and then describe in detail the construction se-uence for this way of making a tank. Read this through ll the way a few times and don't start your tank until nd unless it makes perfect sense. A tank bigger than a ew thousand gallons is way too big of a project to screw p. If you want to make a big tank, but feel uneasy bout following this procedure, I suggest you make a maller one first to get familiar with the process. (See igure 26, p. 73, Plumbing Options for Multiple Tanks.) ılso, I suggest you carefully read the Appendix B, p. 1, Tank Loads and Structural Considerations, as well. specially if you are building a 30,000 gal or bigger tank, is essential that you have a good grasp of the struc-ural considerations before diving in.

Note: This procedure is so geared to US materials, meth-ds, and context, we've elected not to clog the text with metric onversions and materials. Most of the conversions can be und in Appendix A, p. 90.

Many thanks to Paul Kemnitzer,[21] ferrocement tank pioneer, on whose 22 years' experience most of this section is based. He can be reached at (805) 451-5153 or pabloteebs@ gmail.com.

Tools

In approximate order of appearance...

Tape measures—*Measure twice, cut once*
Water level—*For leveling floor*
String—*For marking radius, levels*
Level—*For checking angle on drainpipe, floor, plumb on walls*
Flat shovel, pick—*For site prep*
***Digging bar**—*For site prep, levering the armature up to put the dobies underneath*
PVC saw—*To cut drainpipe*
Diagonal cutters or mini-bolt cutters—*For welded wire mesh*
***Hog ring pliers**—*For hog rings*[37]
***A "Willard" rebar cutter-bender**—*Expensive but very handy. See if you can borrow one. If not, for cutting, use a hacksaw, an angle grinder, or a worm drive saw with a metal blade. For bending, use the next two items below (which are useful even if you do most bending on the Willard).*
Rebar hickey—*For bending rebar*
30" length of ¾" galvanized pipe—*Handy for bending rebar*
Tie wire swivel tool—*To manually tie rebar tie wire*
Linesman's pliers—*To tie off tie wire*
Tin snips—*To cut hardware cloth, lath*
6' and 8' step ladders—*To work on roof and ceiling*
***Pneumatic hog ring pliers**—*(Optional) to install the thousands of wire ties necessary to hold the armature together tight. Hog ring pliers and/or pneumatic hog ring guns can be employed for a huge time savings over doing the whole thing by hand with wire.*
Pool trowels—*One per finisher (other trowel types can be used if that's all you've got)*
Wood saw—*To cut roof braces to length*
***Air compressor**—*(Optional, or CO_2 tanks to run air tools)*

TABLE 12: TANK MATERIALS AND SPECIFICATIONS BY TANK SIZE

Material	ea. cost	5,000 gal	10,000 gal	15,000 gal	20,000 gal	30,000
$^3/_8$" rebar (20'pieces)	$6.52	30	50	60	70	50
½" rebar (20' pieces)	$9.71					50
Lath (27"x8' pieces)	$8.30	27	40	50	65	80
6x6x10x10 welded wire mesh (7'x200'rolls)	$290.00	1	1.25	1.5	2	2
½" Hardware cloth (4'x100' rolls)	$139.00	1	1.75	2	3	3
Tie wire, 6" precut, w/loops on ends (big bundle)	$4.60	2	2	3	3	3
Cement (94 lb bags)	$9.60	18	25	32	35	50
Plaster sand (yd^3)	$46.00	4	4.5	5.5	7	9
Water (gal)	$0.01	500	750	1,000	1,250	1,50
Thoroughseal/Bonsal Sure Coat (50 lb bags)	$32.00	7	10	15	17	20
Color (lbs)	$5.00	5	7	10	12	15
Loose hog rings (25 lb boxes)[37]	$65.00	1	1	1	1	2
Hog ring gun staples (boxes of 10,000)	$65.00	1	2	2	2	3
Dobies	$0.32	50	75	100	120	150
4x4 Poles	$16.50	6	10	15	20	40
Concrete (yd^3)	$262.00	2	3.5	4.5	6	8
Approximate materials cost (2009)		**$2,238**	**$3,383**	**$4,287**	**$5,417**	**$7,23**

Water—*Under pressure, if possible (or ability to set up pump to create pressure)*

Rubber boots—*For slab work*

***Concrete tamper**—*For getting voids out of slab*

***Plaster mixer, pump**—*(Usually comes with a crew of plasterers) or—*

Cement mixer, shovels, buckets, wheelbarrow—*To mix plaster by hand*

Mason's brush—*For applying color coat*

5 gal buckets

Heavy rubber gloves

* *Tools denoted with an asterisk are optional. The other tools are needed for all ferrocement construction techniques.*

Materials

The quantities in Table 12 (above) are approximate. Concrete, especially, is critical. Calculate the amount for your tank shape and slab thickness using the formulas in Appendix A or our Tank Calculator.[6] Always order 10–20% more concrete than you calculate that you will use. Notes on materials:

Dobies—*To maintain spacing between floor rebar and the earth*

Poles—*To support ceiling: 4x4" or 3" round minimum for large tanks*

Pallet wrap—*To wrap tank to slow moisture loss from plaster*

Concrete—*1:2:3 cement: sand: gravel is a good mix; or 6 sacks per cubic yard when ordering by truck*

Plaster—*1:3 cement: sand should be sufficient (Paul uses 1:2 in his tanks)*

Labor

Design, site prep—*Varies*

Armature prep—*Approximate times, assuming easy access, materials already on site, no learning curve delays, no design problems or big mistakes:*

- 2,000 gal tank: three days for two people
- 5,000 gal tank: four days for two people
- 50,000 gal tank: three to four weeks for four people
- 100,000 gal tank: two months for six people

Pouring the floor—*With a concrete truck and pumper is counted as one of the above days*

Plaster—

- with mixer and pump, one day
- by hand, up to 20,000 gal in one day with eight people, two days for larger tanks

Then you wait a week....

Color coat and clean-up—*Two work days separated by a drying day*

Design

Make a detailed drawing of your tank and try as much as possible to work out any design issues on paper.

These instructions should enable a handy person with skilled help on the plastering to make a tank up to 30,000 gal in size. It can be scaled down to 5,000 or 3,000 gal tanks. Any smaller than that and the degree of overbuild is ridiculous—use medium-duty ferrocement construction.

For larger tanks (up to 100,000 gal), you'll want to engage an engineer to test the soil the tank is going to be resting on, and to specify the spacing of rebar and the slab thickness.

Site Prep

Here's a site prep checklist. Access is the first order of business.

- **If you can drive a truck right up to the tank site**—And have water under pressure and power right there, that is ideal.
- **If you can run a plaster pump hose**—From the closest vehicle access to the tank site, that will save carrying the heaviest material.
- **If the site is walk-in only**—You can still do it, but you're going to have to carry everything, as well as mix and apply the plaster by hand (unless you are able to pump the plaster, which can be done 350–500' even uphill).
- **Pump water to a small temporary tank higher up**—If the terrain permits, you can make a temporary pressurized water system for the tank work site.
- **If the site is vegetated**—You'll want to clear an area a few times bigger than the tank, to provide clearance to work, store materials, and mix plaster.
- **If the soil has a high clay content**—You'll want to lay down 6" of compacted road base or gravel to pour the floor on.

Grade for the Floor

Most of these tanks have been built with a flat floor and a drain sump at one side. This is significantly easier to build than a sloped floor, and it is easy enough to sweep the resulting puddles into the sump during tank cleaning with a broom. *(Check the sidebar on p. 112 for experimental floor options that could save material, add strength, and make cleaning easier.)*

For a flat floor, you just cut a flat space into the earth to pour on. How evenly do you need to dig? A tolerance tighter than ½" perfect is a waste of time.

Drain

Now dig a hole for the drain sump, and a trench for the drainpipe.

The drain sump should be a few inches deeper than the low point of the floor, and a foot or two around. However thick the slab is, the excavation will be that much bigger all the way around.

See Drain, p. 58, for more info and complete specs for the drain pipe, including leak-prevention measures: it is critical that the tank not leak where the drainpipe passes through, as there is no access to do a repair.

The drainpipe should slope 2%. The pipe at the drain sump end should terminate in a coupling, a 45°, a 90°, or a straight coupling. (See Figures 40, 41, p. 112, 113.)

You're going to fill the whole trench with concrete, with at least 3" around the pipe on all sides, to about 6" past the tank edge. You can stop the concrete from flowing farther by packing rocks around the pipe. The portion of the pipe that is to be encased in concrete should be held securely up off the trench floor.

To discourage the concrete from cracking, you can wrap the pipe in a cylinder of welded wire mesh. The concrete must adhere perfectly to the pipe for a good seal. If you've got at least 3" of strong concrete with a rebar containing the pressure, you can put a ½" thick ring of bentonite / tar cold joint seal around the pipe. This will expand with great force if water touches it, and make a complete seal (or crack the concrete if it is not as thick as specified). Just before the pipe exits the cistern, it may be helpful to give it a few turns of insulating pipe wrap so it can wiggle in the grip of the cistern and not crack, if the tank shifts under its great weight. The backfill should be free of rocks that could break the pipe if this happens.

Where to Apply Your Perfectionism

Ferrocement is a very flexible medium. It is an inappropriate material with which to attempt the perfect geometric symmetry of a metal or plastic tank. If the tank is slightly oval, or the roof curve variable, or the walls way out of plumb, it's really not going to hurt anything. Ditto for an uneven surface finish. Ferrocement is well suited to an organic, artistic sort of perfectionism, finding the perfectly pleasing asymmetric curve or pleasing natural boulder shape.

However, there is a role for detail orientation in ensuring that the armature is securely tied off everywhere, and that the plaster fills the walls without any voids or cold joints.

Floor and Inside Wall: Welded Wire Mesh

Lay strips of welded wire mesh ("mesh" henceforth) across the entire area of the floor, overlapping the edges between strips one or two squares (6"–12"), and extending beyond the perimeter of the tank at least two squares (see Figure 39, next page). If you've gone with a conical or domed floor, you may need to cut shorter strips of mesh and overlap them. All the joints should be hog ringed together so the mesh sits somewhat flat. It will be strongest (and the least work) with the longest strips that can negotiate the shape.

Now cut a strip of full-height (7') mesh to the length of the circumference of the tank, plus a few feet. Stand this up to make a cylindrical wall, and jockey it into position in the circumference line (6" in from the perimeter stakes), hog ringing it all along its bottom edge to the mesh on the floor.

Now you've got a pretty good indication of what the final tank is going to look like. If you want to change its diameter or location, speak now or forever hold your peace!

You'll also notice that you no longer have a way to get in and out of the tank.

Cut the smallest door you can get through in the mesh. The door should start two squares (1') above the ground and be maybe 2'and 3' high. Some slit polyethylene drip tubing can be placed over the cut edges to reduce snagging and scraping on body parts.

Double rebar tied.

Tying off welded wire mesh.

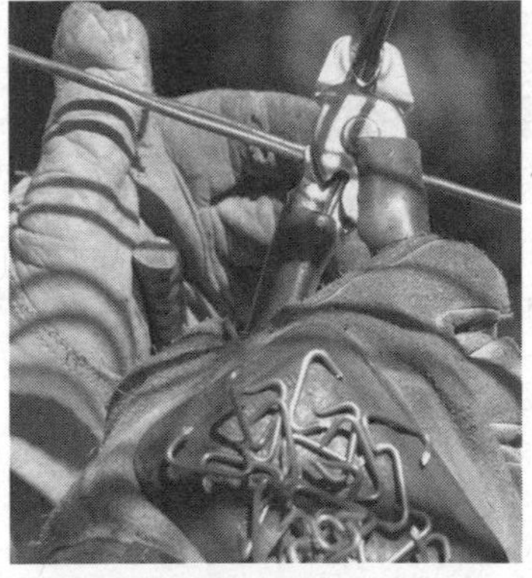
Hog ring pliers and cool magnetic glove loaded with rings.

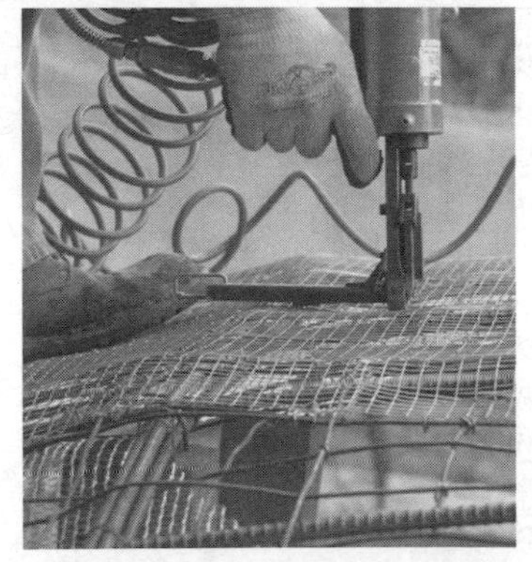
Pneumatic hog ring gun.

CO_2 tank to run pneumatic hog ring pliers without a compressor.

Figure 39: Heavy-Duty Ferrocement Roof and Floor Plans

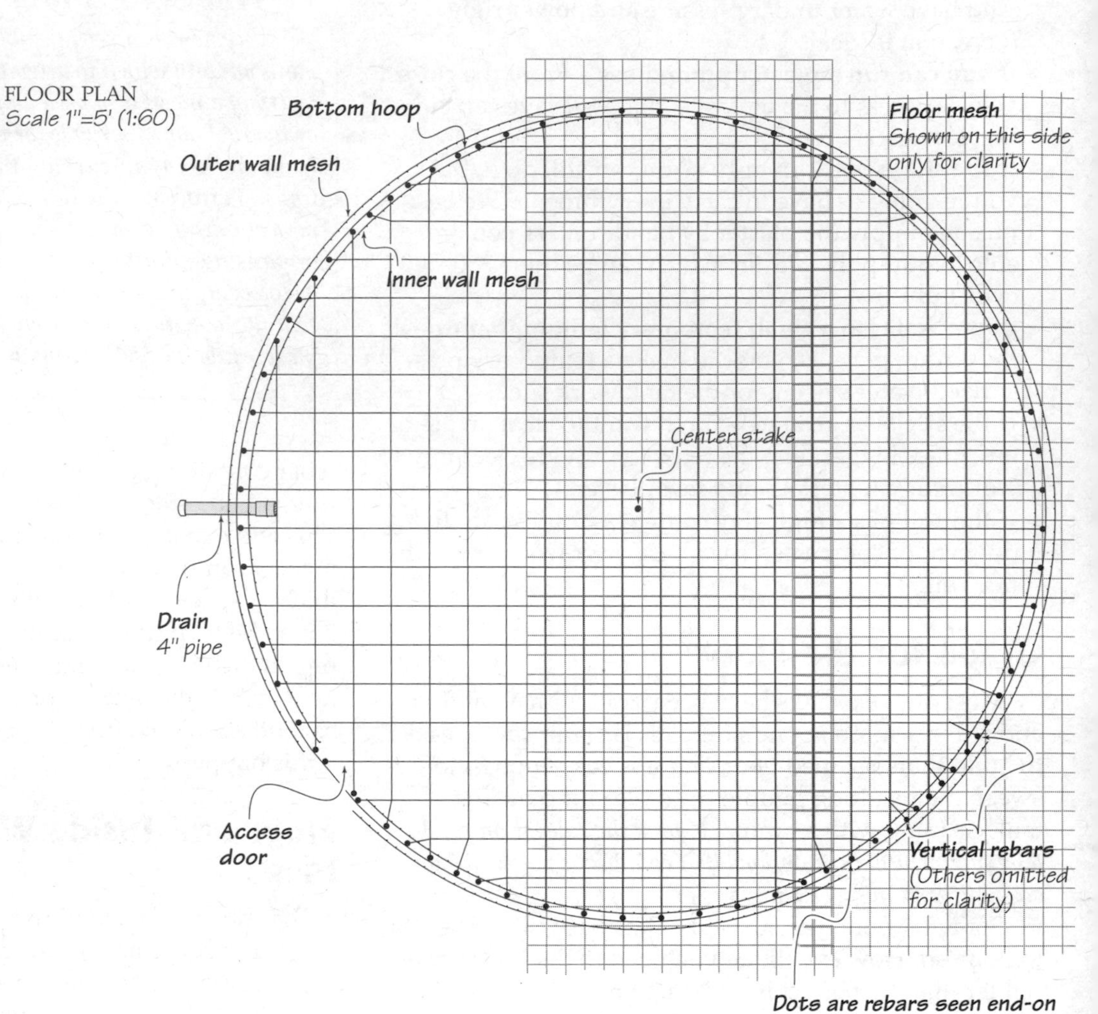

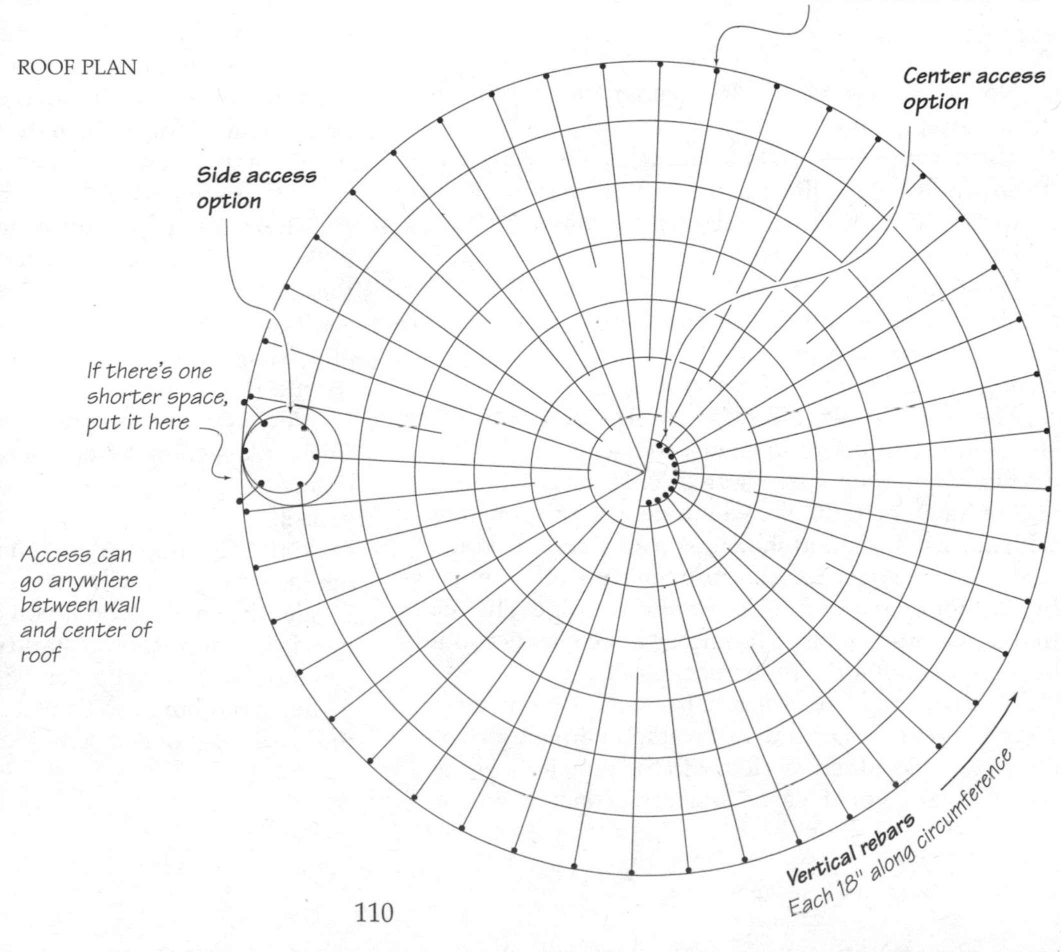

Floor and Wall: Rebar

The floor and roof can be reinforced with rebar in a *grid* or a *radial with hoops* pattern (Figure 39). Because of the different ways roofs and floors are loaded, roofs are stronger with radial rebars and hoops, and floors are stronger with a grid. If the two systems cross at the floor, there will be an extra concentration of rebar ends at the floor to wall joint, where it is most sorely needed. This is the way I prefer to do it for a tank this size, but either pattern can work for either roof or floor. (Note: if you make both floor and roof radial, you can run rebars continuously from the middle of the floor to the middle of the roof, simplifying the construction of small tanks.)

For radial with hoops, bend the rebars into long-footed "L" shapes. Slide the feet of the Ls through the bottom hole in the mesh (so the feet are on top of the floor mesh and the verticals are outside the wall mesh). Now tie them in place to the wall mesh with one double rebar tie, to keep them from flopping over. The first two verticals should go on either side of the door. The rest go along vertical wires on the mesh, every two, three, or four squares, depending on the size of the tank. Every 2' (four squares) has proven to be enough for tanks smaller than 20,000 gal, every 18" for up to 30,000 gal, and every foot for 40,000–100,000 gal.

The feet of the Ls should go to a ring around the drain sump, with just a few long enough to be bent and cross under the drain sump to the ring over on the other side. The vertical pieces can just stick up however far they do.

Experimental Center Drain

For a center drain (which is what you'll have with a conical or round floor, Figure 41) you could either fill the whole trench with concrete, or backfill around the pipe with well-compacted, wetted earth or sand, which will save some concrete, at the cost of some extra diddling. If you do this, you reduce the chance of the pipe cracking, but increase the chance of the slab cracking where it spans the trench—perfect compaction is imperative. If you backfill, leave the first foot of the pipe unburied, so it will end up encased in concrete for a good seal. You can hold the pipe in position by wiring it to the rebar with wired-on 3" dobies as spacers between both the rebar and the earth below.

Tying off expanding metal lath on roof with pneumatic hog ring pliers.

FIGURE 40: HEAVY-DUTY FERROCEMENT TANK (SECTION)

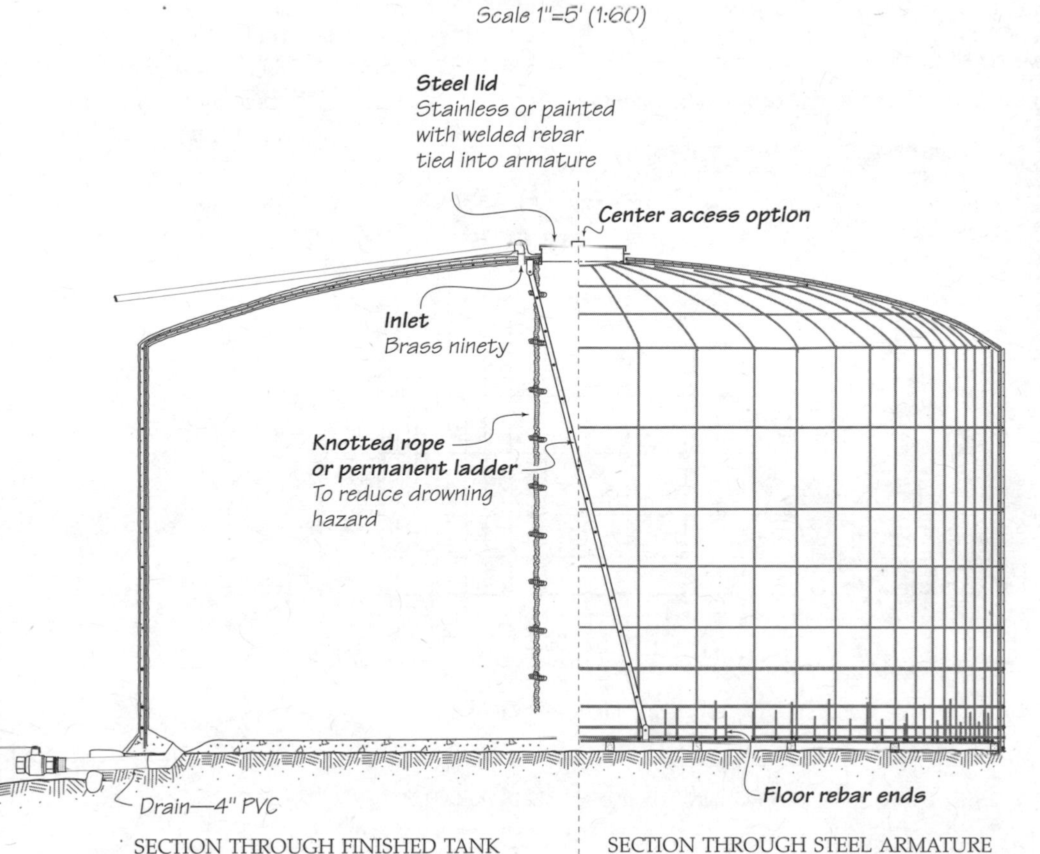

Tying off welded wire mesh with hog ring pliers.

Tempting Floor Shape Innovations

Read this section only if you want to innovate...

The walls and roof get their strength from their curved shape. The floor, if it is flat, gets its strength by brute force of being thick (see Appendix B, Tank Loads). A flat slab floor is as inefficient structurally as the curved walls and compound curved roof are efficient and elegant. Suppose you were to make the floor conical, with a center drain, or compound curved like the dome roof. Or—the bottom of a glass bottle? In theory you could reduce the slab thickness dramatically.

Why isn't this done? Is it simply because it is traditional to pour slabs flat on the ground, because it is much more work to dig a dome-shaped hole?

I don't know. Large, buried hemispherical tanks have been made successfully, and many swimming pools have compound curve floors.

***Conical floor**—A slightly cone-shaped slab, with concrete thickness at the low end of the traditional range promises better draining and improved strength with less material. This hasn't been done that I know of, but it is **very** close to things that have been. This shape should save considerably on concrete, add strength, and not be too hard to build. You could get this shape by pivoting a straight edge from the center along the circumference. It has the advantage over a floor sloped to one side that the walls are of uniform height. (Note that changing the floor changes the drain, too; see sidebar on p. 111.)*

***Dished floor**—Or, go all the way and curve the floor like a dome roof or the bottom of a glass bottle, and make it 2-3" thick instead of 6" or 8"* (Figure 41).

I've used these floors with great success in unreinforced stucco (sand: cement) jar tanks of 160 gal (600 L). *We slope the whole bottom so that the gutter around the perimeter flows toward the drain. We also have the outlet at the high point of the island in the middle, where it is least likely to suck crud.*

This should scale up fine to a large tank. Doubtless it would be more work than a flat floor, but much more economical of material and possibly less likely to leak.

Domed floor in jar tank.

I suggest you add a second layer of welded wire mesh and possibly some wire fencing with 2" squares if you do this, so it has some fine-scale reinforcing like the walls. You'll need an accurate, smooth, well-compacted excavation for a floor so thin.

This shape could be achieved by rotating a curved guide around a tilted pole in the center.

Warning: This drawing is the best composite of several designs and has not been proven in construction. If you try an innovative floor shape, please take good photos and let us know how it works out.

FIGURE 41: POSSIBLE FERROCEMENT DESIGN INNOVATIONS

Steel lid
Stainless or painted with welded rebar tied into armature
Scale 1"=5' (1:60)
Ferrocement lid
Side access option
Center access option
Overflow/vent
Brass coupling
Inlet
Brass ninety
Side access can be flush with the dome (more beautiful), fully telescoped (more functional), or anywhere in between
Permanent ladder
Drain extension
Keeps concrete out of drain
Built-in ladder
Domed floor with perimeter gutter sloped toward drain
Floor rebar ends
Drain
Center stake
String line

SECTION THROUGH FINISHED TANK

SECTION THROUGH STEEL ARMATURE
For clarity, rebar is only shown on back of tank

For grid, the preferred method and the one in the figures, place rebars in a grid, every foot or two, depending on the tank diameter. The ends should protrude beyond the walls 6" to 2'; these get bent up 90° into the plane of the wall. If you do a grid, some of these bent-up wild ends will be where the verticals go, others won't. If you can bend them a bit to get them spaced evenly, that's an advantage, but don't worry about it too much (Figure 39).

This floor-to-wall joint (Figure 42, below) is where the advantage of the grid is. It uses roughly the same amount of material as a radial pattern, but instead of the "extra" rebar density being in the middle of the floor (as is the case with the radial pattern), it is at the wall-to-floor joint, where shear stress is greatest and reinforcement is most needed. These extra rebar ends help keep the walls from tearing outward from the floor under pressure. Bend these wild grid ends up into the plane of the wall.

If you've done the grid, now you'll add verticals that have a short-footed L shape. The short foot of the L should extend into the plane of the floor a foot or two, and be tied to one or more floor rebars. How high should the ends go up? All that's left—the wild ends will get bent down to make the roof. This is where the extra work of the grid floor is: dealing with these wild ends. It may be helpful to put up a hoop of rebar at the top of the mesh to help hold them. Just make sure the verticals end up on the outside of the mesh but inside the hoop.

The horizontal wall rebars (hoops) should be bigger and/or spaced more closely where the loads are greatest. At bottom, both hoop stress (from pressure pushing out) and shear (from the walls trying to push away from the floor) are greatest. Figure 43 shows suggested hoop spacing for different wall heights.

The first hoop is at the very bottom. It holds in all the verticals and bent-up floor ends. The second hoop to be placed is the one which goes 1' up, at the bottom edge of the access door you are about to cut.

All the rebars which

Strong Lower Walls and Wall-to-Floor Joint

Forces in a water tank concentrate at the base of the walls and the wall-to-floor joint. Here are three strategies to resist them:

- *Decrease the rebar spacing and increase rebar size near the bottom of the wall.*
- *Increase the thickness of the wall near the bottom. (Using fatter rebar automatically increases the thickness of the plaster inside the armature, and the plasterers can be instructed to make the coverage thicker near the bottom.)*
- *Fatten the joint between the floor and wall inside and out with a concave fillet of plaster that partially fills the corner and tapers gradually on the wall and floor on both sides.*

FIGURE 42: WALL AND FLOOR JOINT DETAIL, WITH CONSTRUCTION SEQUENCE
Scale 1:4

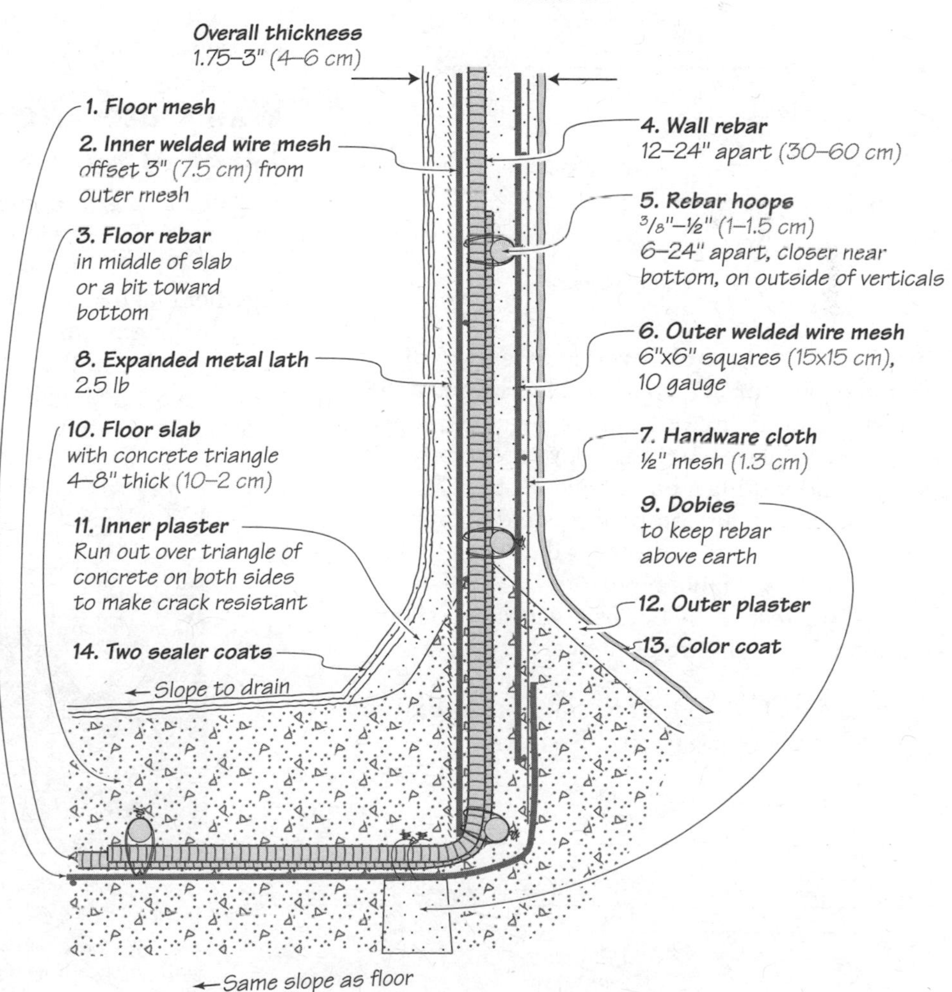

Figure 43: Hoop Spacing for Different Height Walls

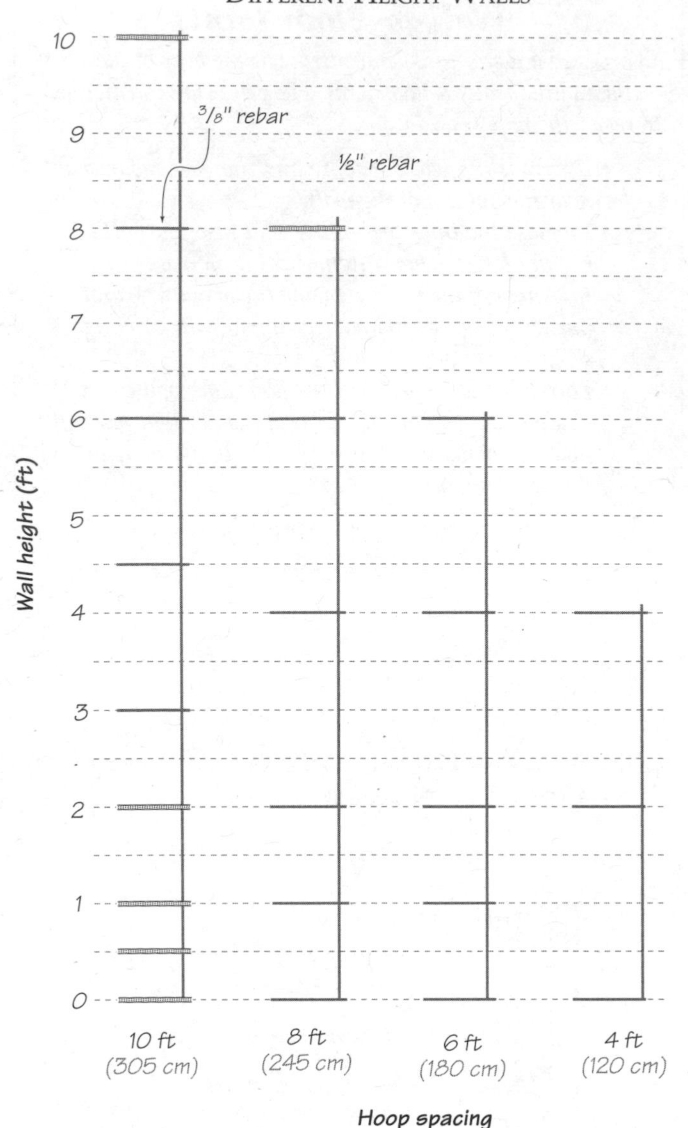

are joined end to end should overlap at least 50 diameters (2' for ½", 18" for 3/8" rebar) and have three tight, double ties to hold them together. This can be fudged a bit in non-critical places, but the hoops are totally critical; they are under quite a bit of tension stress.

By the way, if you are tempted to weld the rebars together, forget it. Welding destroys the strength of rebar (unless it is a rare weldable type).

Now add the rest of the hoops. Every hoop should be tied with a doubled wire tie at every crossing of a

Figure 44: How to Join Rebar End to End for full strength in tension

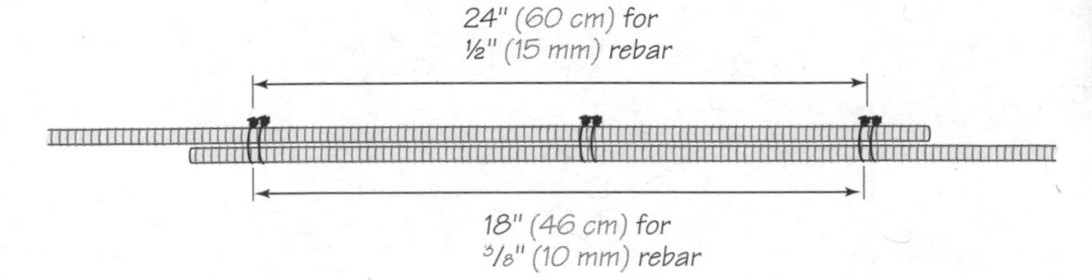

vertical rebar.

If you did the radial floor, now is the time to add the floor circles. Radial or grid, every crossing of floor rebar—every crossing of rebar, period—should be securely hand-tied with a doubled wire tie.

And the mesh should be wired to the rebars all over, so that when you step through a puddle of concrete onto the mesh it doesn't pull loose and get pushed down into the dirt, where it will rust and possibly crack your slab.

Inlet, Outlet, and Overflow Hardware

Before the second layer of mesh is put on is a good time to add inlet, outlet, and overflow fittings. These can be galvanized, PVC, or brass. If there is any chance you will be the person who is going to have to replace the fittings when they rust out, I suggest you use brass. Why build a tank that could last 100 years with fittings that are guaranteed to fail in 30? Brass fittings are expensive, but a bargain compared to the work of replacing rusted-out galvanized ones.

You can weld (or solder, in the case of brass) tangs sticking off of the couplings to keep them from spinning in the wall when someone is trying to crank out a rusted fitting with a pipe wrench (another reason to use brass). The tangs get tied to the rebar. An alternative to welding tangs to give fittings "tooth" is to use 45° bends or tees with a plug in them—the tee is like one big tooth. (See Figure 18, Drain Options, p. 61.)

Wall Outer Welded Wire Mesh, First Layer of Lath and Hardware Cloth

Cut 10' lengths of mesh, and hog ring them to the inner mesh. These pieces of mesh should be offset 3" up and 3" to the side from the inner mesh, so the two together make 3" square holes instead of 6" square holes.

Why lengths just 10' long? Since the outer mesh is on the outside of the wall rebar, it is making a circle that is about an inch bigger in radius. Thus, the longer the sections, the further out of sync it will get with the verti-

Close-up of wall armature; ½" rebar on 1' centers, two layers of 6" x 6" mesh, offset to make 3" squares, sandwiched between expanded metal lath on one side and chicken wire on the other.

cal rebars and the inner mesh. Ten feet seems to be the longest that works for keeping the wires somewhat in sync.

The two layers of mesh should be hog ringed to each other all over the place. Push with your hand. Do they separate? Hog ring 'em. You can use a short piece of rebar to pry the pieces together for joining. In a 2' x 2' section between rebars, it is not unreasonable to have ten hog rings.

Those wings of excess floor mesh which you've been tripping on this whole time? Now is the time to bend them up out of the way. You can cut off the excess on corners, bending them up and hog ringing them to the outside mesh. As mentioned before, the more steel and the thicker the plaster near the bottom the better.

Add the bottom rows of expanded metal lath ("lath" henceforth) inside and hardware cloth on the outside of the wall. These will get buried a few inches in the concrete of the floor and help take the shear force between the walls and floor. These two items get stapled with the pneumatic hog ring gun if you've got the fancy tool, or with the same old manual hog rings if not.

Pumping and spreading concrete for the floor.

Lift the Whole Thing Up and Get Ready to Pour

Get out your digging bar and lever the armature up onto dobies, starting at one side and working your way toward the other. (You can see dobies in action in the photo above.)

How big should the dobies be? Well, the rebar should end up in the middle or slightly toward the bottom of the slab for greatest strength and protection against corrosion.

I'd go with 2" dobies for a 5" slab, 3" for a 6" or 8" slab. If you are making a small tank (10' diameter or less) and want to spend extra time to save material, grade the earth very perfectly and use 1.5" dobies and a 4" slab. With small dobies, take care they don't sink into the earth too much (or use 2" dobies and stomp them into the earth ½").

The dobies should be spaced such that you can walk on the rebar without it bending so much that the ties loosen—every couple to four feet, depending on the stiffness of the rebar.

Check around that all is ready.

Plug the drain line at the sump so it doesn't fill with concrete, and tape the openings of the outlets so the threads don't foul with splatters of concrete.

Check that there is 3" of clearance all around the drainpipe for a good seal with concrete. Pull out any blown leaves or fallen dirt from around the drain. Place securely wired scraps of rebar or dobies between the floor rebar and the pipe to push it down against its supporting dobies, so the pressure of inrushing concrete from the hose cannot move it in any direction.

Pour the Floor

The easy, expensive way to do the pour is to hire a truck to mix and bring the concrete, a pumper to spread it evenly over the floor, and a finisher or two to smooth and level it. In this case you just ask for their strong mix (six 94 lb sacks of cement per cubic yard) with pea gravel, so it can make it through the pumper hose.

The hard way is to get a bunch of your friends to mix the cement by hand (or mixer), carry it in through that little door in wheelbarrows or buckets, and finish it yourself. In this case you can use a 1:2:3 cement: sand: gravel mix. ¾" gravel will add more crack-resistance.

If you've ever poured concrete, I don't need to tell you that you want everything tidied up, all tools and materials on hand, and all hands on deck for an early start on pouring day. It is *imperative* that the slab for a water tank be completed in one continuous pour, with no cold joints. If it cracks, it's not just an aesthetic issue; it is a big leak at the very bottom of your tank.

With a truck, pumper, and two finishers, you should be able to finish a slab up to 20' across before the concrete sets up too much, i.e., half a day. You might want to add a few more helpers for a bigger-diameter slab.

I would not even attempt a hand-mixed, hand-poured water tank slab any bigger than 6' across and 4" thick without a generously sized, proven crew that you are positive—based on past performance—can finish up before dark. And I'd still have klieg lights on hand just in case....

If the soil is dry, wet lightly before the concrete goes on.

As the concrete is applied around the edges, someone with rubber boots needs to spooge it to the outside several inches, as someone outside pushes it up three to several inches on the outside of the wall armature. The ideal is to end up with sort of a triangular fillet of concrete on both sides of the wall—see Figure 42, p. 113. This triangle will be encased with wall plaster on both sides, providing a much less crack- and leak-prone joint to the wall than a simple butt joint. Of course, it's best to work from the farthest reaches of the inside back toward the door.

At critical places, such as around (and especially under) the drain line, you can ensure a void-free fill by filling from the bottom up. If you shove the hose down next to the (well-anchored) pipe, concrete will boil up on both sides of it from the bottom up, pushing air ahead of it. You can also push around it with a trowel.

Tamp the concrete into place with a concrete tamper to ensure that it is void-free. (You can see one in the photo on p. 115.)

The floor slope can be judged by eye and finished fairly rough, but draining the tank may then leave puddles. This will be less work to make, more work to clean.

If, on the other hand, the idea of an easily made yet wobbly floor offends thee, enlist the aid of a screed setup designed by someone who has experience making slabs to get the slope just right (see sidebar at right).

Once the floor has set up hard enough to kneel on atop a small piece of plywood without sinking in, it is time to smooth out the top surface. Impress on your helpers that you do not want the mirror-smooth, shuffleboard-court type finish it seems you would want for easy cleaning. You want to leave enough "tooth" that the Thoroughseal coat sticks. If you want a smooth finish (which isn't as crucial as slope for cleaning ease, since you'll be sweeping it anyway), you'll have to try and make it in the sealer coats.

The drain sump is a good thing to have smooth, as this is where all the gunk will congregate and resist going down the drain. Make sure no part of the sump is below the spill point of the drain.

After you've convinced yourself the pour has been finished perfectly, clean the tools and go to sleep.

The next morning you can securely cap the drain, flood the slab (up to the top of the little ridge of concrete you pushed up around the perimeter), and forget about it for a day or three while you pay attention to your friends and family. Stop by once in a while and spray down the outside edge of the slab, especially on the south side.

You can slow the drying by putting wet blankets over the concrete and plastic tarps over that. The slower it dries, the less likely it is to crack. You should keep it sopping wet for a couple of days and damp for a week or more.

Finishing of Conical or Flat Floors

For a conical floor, pivot a string about a removable stake or brick in the center as a guide for finishing. With the 4% slope specified in the drawing of this optional design innovation (p. 112), this should be sufficient. After the string has done its job, pull the stake out of the middle of the still mushy slab, and stir up the concrete to fill the hole perfectly.

For flatter slopes, it should be possible to screed off the floor from a center pivot to something fastened around the circumference to get a nearly perfect cone that will drain puddle-free.

To finish a floor that slopes to one point on the side, you can set wooden guides on the rebar whose surfaces are at finished floor level. Pour in between them and screed over the top of them. Then, right after you've got the rest of the floor level, pull them out and patch the holes, while the concrete has yet to set up enough to do the final finish.

The Roof, Cool Shapes, Ladder

Now that the rebars are held solidly at the bottom in cured concrete, you can bend down the wild ends of the vertical rebars to form the roof.

How domed a roof should you make? The more curved it is, the harder it is to make (because of the awkward fit of the mesh to the compound curve). A high, steeply curved dome makes a stronger roof. However, this advantage is outweighed by the large increase in water pressure on the lower wall for a relative pittance in increased storage capacity.

I advocate a fairly flat dome, rising about $^1/_{10}$th of its diameter. The rings on a radial pattern roof contain the strong outward forces generated as people walk on it. That is the structural advantage of the radial pattern for a roof. On a big roof, it is a nice touch to make the top wall hoop a rebar-size fatter, or add another near it for good measure. (Remember, all the hoops go outside the verticals, to contain them.) Rebars joined end-to-end should be overlapped at least 50 diameters, especially the rings and hoops, which are under tension. (See Figure 44, p. 115.)

What curve should you use? The choice is not critical, unless you are making a flat dome, in which case the strongest curve overall is a section of circle, that is, a constant curve. If the curve is variable, for instance, more curved near the wall and less curved in the middle (which is what you'll get if you just pull the rebar down with your hand), the less curved section will be less able to resist point loads from people walking on it.

If you want "wings" or a lip on the side of the tank so that it can harvest water from its own roof, now is the time to add them. Wings will require another hoop, held in place by short verticals or the ends of roof grid members that extend beyond the edge. There is little load on modest-sized (under 2') wings, so they can be lightly reinforced. You can make them thin, with a fat lip of lath to strengthen the outside edge.

The vertical rebars, which are spaced evenly around the circumference, can be tweaked over to line up with roof grid lines, if you're doing a grid roof.

As it is under the least pressure, the roof isn't so critical. If the grid is a bit—or a lot—wank-a-doodle, it will still be fine.

This is the time to attach the access hatch, if you've had one made, or form one if you're making it yourself. An access hatch on the edge has the advantage that it can be located over an integrated ladder in the wall, and you'll have an easier time getting in to plaster the inside without trashing the fresh plaster on the roof.

If you telescope it up, as shown in Figure 41, p. 112, it gives you a point to attach your inlet above the high water level without snaking the pipe over your lovely roof. (Or, if you really care about the aesthetics, pass the inlet through low on the wall of the tank and telescope up above the maximum water level inside the tank, a measure which can help against freezing, too.)

An access hatch at the high point, in the middle (or telescoped up), has the advantage that it can potentially enable quite a bit more water to be stored above the walls but below the roof.

If you want to make a rock-shaped tank, this is the time to get creative.

In order to resist the water pressure, all parts of the walls that have 4' or more of water pressing on them need to be circular. That is, the bottom four feet of an 8' tank, the bottom 6' of a 10' tank. The closer you get to the top, the lower the water pressure and the more wild you can get with the shape, without inviting structural problems. Compound curves are inherently strong, and can start lower on the tank wall, and changes in diameter (as in the urn-shaped tanks described previously) are still circular, and thus don't compromise strength. The stress on the roof is low. You can make it all bumpy and asymmetrical, starting at different heights, and it will still be plenty strong, as long as there are not big flat sections (see photo, p. 38).

The pressure at the bottom of the tank is proportional to the height of the highest water level, regardless of how much volume of water is up there. This is a species of leverage. If you make a hollow "rock" with a narrow spike sticking up 20', even if the spike is only a few inches around, if the thing can fill with water it will add 20' of pressure to the tank. This might well blow apart even a massive lower structure (see Figure 32, p. 91).

Conversely, fanning out to a wider diameter above (as in an urn) adds no more pressure below. This is why clay urns are urn-shaped; the flare at top adds volume without increasing the tension on the clay at the bottom.

For a regular domed, radial rebar roof, you just give the wall rebars a concentrated bend at the top of the wall, directly down almost 90°, with a pipe and rebar hickey, until they are in the plane of the dome edge. From there you can give them a gentle, distributed bend into the rest of the dome curve.

Once you've got the rebar all tied off, drape strips of mesh over the whole top, with the usual 1 to 2-square overlap. These need to be hog ringed where they are sticking up above the rebar. If the dome is very curved, you may need to clip the mesh in places (or twist individual wires into "Z" shapes to shorten them) to get the mesh to conform to the curve.

The ceiling welded-wire mesh is installed in similar fashion, except that it is made of shorter (10') strips, tightly rolled and fed through the side door. These must be well hog ringed, as they may find themselves supporting a big slug of heavy, wet plaster and a mason's foot.

If you are going to make an integrated ladder, this is the time to put it in. For specs, see Service Access, p. 56.

If you want, you can add a pole or permanent ladder in the middle. If the ladder is steel, you can bolt it to anchors in the floor slab, let it slide on the inside of the hatch during construction, then cut it to length and bolt or weld it to the hatch after the tank has cured (the tank will shrink). If you use a steel pole, 2" is plenty. It should be bolted to the floor, and covered with hardware cloth and plastered to keep it from corroding.

Lath and Hardware Cloth

Lath goes on the inside surface of the whole thing now. Use 2.5 lb/yd^2 galvanized lath, with the cups facing up, and edges overlapping 2". Lath goes on the inside to catch the plaster you apply from the outside.

Plaster easily passes through the ½" hardware cloth, with which you are now going to skin the entire outside. ½" chicken wire is cheaper and can also be used, but hardware cloth is stronger and smoother to trowel over. (Chicken wire or lightest-weight expanded metal lath can conform to compound curves and are the materials of choice for any small, curvy details you've incorporated.)

Both lath and hardware cloth can be attached with the pneumatic hog ring gun, if you're so lucky as to have one. Both should be checked all over by pushing with your hand, and hog ringed really tight. Twenty hog rings in a 2' x 2' square are not excessive. The lath on the inside of the ceiling should be especially tight. If it pushes in as the cement is going on, it's not going to be possible to pull it back once it's filled with cement.

Plaster Prep: Roof Supports, Seal Door

The plaster doesn't contribute much to the strength of the armature until it has dried for several hours. Since the tank is plastered from the top down, the armature itself must be made stiff and strong enough to support much of the weight of a few workers and tons of wet plaster, with the help of numerous supports under the roof.

For tanks bigger than 10,000 gal, a mixture of 4 x 4s and 2 x 6s will do as roof supports. For smaller tanks, 2 x 4s or 2 x 6s will work. Undressed round, green poles also work fine.

If you drive a nail through the lath into the end of the poles, that will keep them from falling over. If the supports lean ever so slightly to one side they will be easier to knock out afterward.

Seal off the temporary access door now, patching over it with mesh, lath, and hardware cloth. Be sure to overlap everything generously.

Clean off the slab (so you can scoop up and reuse plaster that lands on it).

Clean any debris out of the tangle of rebar and mesh at the floor-to-wall joint with compressed air or water. You can also blowtorch any trapped leaves. It is crucial that this area be clean for a leak-free joint.

Check that all is tied off perfectly, have bulk sand and cement delivered, and arrange for plenty of help. If you go the plaster mixer route, you'll want:

- a big plaster mixer and a plaster pump
- one semi-skilled person to make the mix
- one person to direct the hose end
- two skilled plasterers to trowel out the plaster as it goes on
- one laborer to move hoses, clean tools, etc.

These figures include whatever workers come with the equipment. As mentioned before, you'll want to start early. With a good crew, you can mostly just watch them go, and be finished before you get too hungry. For example, a five-person crew recently finished a 10,000 gal tank in five hours.

If you are manually mixing the plaster, you'll want:

- two semi-skilled people to mix plaster
- one or two laborers to deliver it to the finishers
- two to four skilled finishers to apply and finish the plaster as it goes on
- one laborer to clean tools, etc.

It is all but impossible to finish an entire big tank manually in a day. For example, a good crew of four was able to do a 45,000 gal tank in two long, hard days.

For mixer or by hand, the plaster is really rich: one part cement to two or three parts sand. 1:3 is OK if the plaster sticks well enough.

Don't Cave the Roof In—A Cautionary Tale

This bracing bent, then snapped, with the result that half the roof caved in.

The braces were long 2 x 6s, which bent in the 2" dimension, then broke. 4 x 4s probably would not have broken.

A cubic yard of cement for the roof was pumped into a big pile in one spot on the roof, together with eight guys to help spread it. Had the cement been distributed with the hose over the whole roof as it came out, the failure probably wouldn't have occurred.

Plaster the Whole Tank

Sequence your work to avoid disturbing fresh plaster. If you start from the bottom and work up, you'll be leaning ladders against wet plaster. For this reason, the preferred sequence is to start at the top and work down. Do the area around the access hole first, and it will be better cured when you've finished a phase and have to go in or out of it.

If you are using a plaster pump, first spray inside, from top to bottom, with a very, very light coat ($^1/_8$"). Plaster comes out the pump hose sort of like popping popcorn. If the hose is held back a ways, it will result in lighter coverage as the spray covers a wider area. Let it cure for half an hour. This coat keeps the heavy plaster coat that is coming next from pouring through the inside.

Now spray the outside to a thickness that fills the armature completely to a voidless condition and leaves a generous, protecting cover of ½" to ¾" over it. After you've covered the entire outside and troweled it to its final finish, take a break, and eat some food while the plaster is setting up a bit, before heading inside.

If the access is in the middle, run a plank from a stepladder outside the tank to the well-supported lip of the access hatch. Run the plaster hose along the plank. It's going to be dark inside now, and hot, from the chemical reaction of the cement. You may need lights to do a good job. You could run an extension through the inlet. If plaster has rained onto the floor from the outside, you can use it for the fillet between the floor and walls.

Give the inside a medium coat ($^3/_8$" to $^5/_8$" cover over the armature—up to 1" if sprayed) from top to bottom, troweled to the final finish.

If you are manually mixing the plaster, the plaster is

PAUL KEMNITZER

40,000 gal (150 m³) ferrocement tank, ready for plaster.

applied from top to bottom, as with the plaster pump. With hand mixing, it is better to apply and finish the plaster on the inside and outside in the same place at the same time, with one finisher on the inside and one on the outside.

After all the plastering is done, clean the tools and get some sleep.

PAUL KEMNITZER

A 100,000 gal (380 m^3) tank ready for plaster.

Keep It Wet

As soon as the plaster gets hard, start wetting the tank down.

One tank maker in Hawaii wraps his tanks in pallet wrap. Since the water then can't escape the tank, it needs little attention. Plug the drain and flood the cured floor with water to keep it humid, and you'll get a really nice, slow cure. Alternatively, you can just keep wetting the plaster regularly, with blankets and tarps to store water and keep it from evaporating rapidly. Some people have rigged automatic sprinkler setups.

After three days, you can remove the supports from the inside, clipping off the protruding nails which held them in place. The holes where the supports were will need patching.

Any remaining drips that landed on the floor can be chipped off now.

Keep the tank saturated wet for three days and damp for a week or two, especially on the sunny side.

Color and Seal It

A week later, after the cement has cured fully, you can add the color coat to the outside. The color goes into the Thoroughseal. (Thoroughseal is a cementatious sealer. Thoroughseal "foundation coat" is cheaper and, being grey instead of white, makes more attractive earth tones.) At the same time the inside can be sealed with two coats of Thoroughseal.

You may be tempted to use 1:1 sand and cement instead of Thoroughseal. It is certainly cheaper, but not as durable. On a related note, silica sand tends to pop out of the wall. Whatever sealer you use inside, you should confirm that it is safe for potable water.[20] Adding some color to the first interior seal coat (only) makes it easier to confirm coverage of both coats inside the dark tank. The Thoroughseal should be about the consistency of pancake batter. It's applied with a big mason's brush.

The color is a place where you can get really creative. If you are particular about the color, try it on big swatches, keeping very careful track of the measurements and procedure. (Davis concrete pigments make the best plaster colors.)

All sorts of things affect how the color turns out; the sand, the cement, the amount of water, and the curing conditions... I suggest you forget about getting a uniform color, and try instead for an interesting patina.

Besides mixing the color into the cement, you can also sprinkle it onto the surface and work it in to get a more mottled appearance. Note that it must be stirred in well—if there are isolated clumps of concentrated color, they are likely to degrade rapidly. Bear in mind that colors lighten dramatically over time. Red, for some reason, seems more durable than other colors, so things tend to turn out more pink than intended.

Once the color is done, you can backfill around the tank and do the finish grading. The ground around the tank should be mounded so that runoff will head away from the tank.

Fill It

A new ferrocement tank should be kept mostly full for a month or two for the final curing. It is normal for a new tank to weep slight amounts of water in places (often on the sunny side, where the cure was too fast). Don't be alarmed—usually this seals up.

As with a cement swimming pool, it is best for the tank if you never let it dry out completely.

There you go—you've got a lifetime water tank!

A completed 40,000 gal (150 m^3) tank.

PAUL KEMNITZER

Endnotes

Updates & clickable links: oasisdesign.net/water/storage

[1] **Principles of Ecological Design** Art Ludwig. Oasis Design. *Principles for redesigning our way of life to live better with less resource use. See description on inside back cover and free online summary at* www.oasisdesign.net/design/principles.htm.

[2] **Create an Oasis with Greywater** Art Ludwig. Oasis Design. *See inside back cover. There is a free online summary of common greywater mistakes and preferred practices at* www.oasisdesign.net/greywater/misinfo.

[3] **Rainwater Harvesting and Runoff Management** Art Ludwig. Oasis Design. *Forthcoming—see description on inside back cover and* www.oasisdesign.net/water/rainharvesting.

[4] **Water Quality Testing Procedures and Information Packet** Art Ludwig. Oasis Design. *A set of downloadable files which will help you learn water testing techniques and interpret results. See inside back cover, also* www.oasisdesign.net/water/quality/coliform.htm.

[5] **American Journal of Public Health** Robert D. Morris. 1992.

[6] **Water Storage Extras** Oasis Design. *Includes Water Tank Calculator, research notes on materials leaching, bacterial regrowth, disinfection byproducts, permeation, and water system component spreadsheet. See inside back cover and* www.oasisdesign.net/water/storage.

[7] **NRDC's March 1999 petition to the FDA** *Includes report on the results of their four-year study of the bottled water industry, including bacterial and chemical contamination problems. The petition and report find major gaps in bottled water regulation and conclude that bottled water is not necessarily safer than tapwater. See* www.nrdc.org/water/drinking/bw/bwinx.asp.

[8] **National Testing Laboratories, Ltd.** Phone 440-449-2525 or 800-458-3330, Fax 440-449-8585, ntlsales@ntllabs.com. www.ntllabs.com. *Offer a $137 water test for 75 parameters. As detection limits are disappointingly high on their standard tests you are unlikely to find anything unless the water is really bad, but the price can't be beat. They tell us they offer tests with lower limits now, as well.*

[9] **Builder's Greywater Guide** Art Ludwig. Oasis Design. www.oasisdesign.net/greywater. Figure 7, p. 43: How Treatment Capacity and Contamination Plumes Change with Location of Wastewater Application.

[10] **Ponds—Planning, Design, Construction** The USDA Natural Resources Conservation Service (NRCS), Agriculture Handbook 590. *Call your local Natural Resources Conservation Service office to get a copy.*

[11] **Building a Pond** Land Owner Resource Centre. P.O. Box 599, 3889 Rideau Valley Dr., Manotick, Ontario K4M 1A5. 613-692-3571 or 1-800-267-3504, Fax 613-692-0831, info@lrconline.com, www.lrconline.com.

[12] **Pond Construction: Some Practical Considerations** Virginia Cooperative Extension, Fisheries and Wildlife, 1997 PUBLICATION 420-011. www.ext.vt.edu/pubs/fisheries/420-011/420-011.html.

[13] **Local weather station data**. *Try searching the web for "evapotranspiration" and "weather station" and the name of your area. Here's an example of what you're looking for:* wwwcimis.water.ca.gov

[14] **Solutions to Common Fish Pond Problems** L. A. Helfrich, Extension Specialist. Fisheries Virginia Tech. Publication Number 420-019, 1999.

[15] **Natural Swimming Pools/Ponds - The Total Guide**. Total Habitat. www.totalhabitat.com/P&P.html. *Designers and builders of natural swimming pools/ponds.*

[16] **A Handbook of Gravity Flow Water Systems** Thomas D. Jordan, Jr. Intermediate Technology Development Group. 1980. *Portions of* Sizing Water Tanks *section and comments on stone tank shape are paraphrased from this work. For stone tanks they suggest octagonal tanks for diameters less than 2.5 m, hexagonal shape for tanks of at least 2 m, and square tanks for small capacities. Includes procedures for welding HDPE water lines.*

[17] **Effects of Water Age on Distribution System Water Quality** American Water Works Association. *Figure adapted by permission.*

[18] **Rainwater Catchment Systems for Domestic Supply** John Gould and Erik Nissen-Petersen. Intermediate Technology Publications, 1999. *Tank sizing figure adapted by permission.*

[19] **Water Distribution Systems Handbook** Larry W. Mays. American Water Works Association, 1999.

[20] **National Sanitation Foundation** *Searchable database of NSF 61 certified materials and components:* www.nsf.org/Certified/PwsComponents.

[21] **Paul Kemnitzer**, Hollister Ranch 51, Gaviota, CA 93117. 805 451-5153. pabloteebs@gmail.com. *Ferrocement pioneer, and builder of quality ferrocement water tanks, homes and structures in Southern California since 1982. Tanks from 3,000–100,000 gal (11–380 m³), cylindrical or boulder-shaped, in natural colors.*

[22] **Pacific Gunite**, Box 421, Mountain View, HI 96771. 808-968-6059, Fax 808-968-8668, www.pacificgunite.com. *Cylindrical tanks 1,000–50,000+ gal.*

[23] **Report of the New South Wales Chief Health Officer, 1997** www.health.nsw.gov.au/public-health/chorep97/env_watalum.htm.

[24] **Maritime Teak Deck Caulk**

[25] **Water Bladders in Culverts** Earthwrights Designs. 505-986-1719, ezentrix@aol.com. *I suggest you avail yourself of Earthwright's experience if you want one of these. They use aluminized steel culverts with 30 mil PVC inside geotextile, or PP bladders. Cost is about $1/gal installed, capacity 10–50k gal. There is limited access for cleaning. First ones installed in 2002. See* www.waterstructures.com *for other storage bladder examples.*

[26] **Zodiac aboveground pools** www.zodiacpools.com.

[27] **Maruata en el Cruce de Caminos** Art Ludwig. Oasis Design. *Ecological systems designs for an indigenous community in Mexico, including water supply and sanitation. See* www.oasisdesign.net/design/examples/maruata/book.htm.

[28] **Oxidation of Iron and Manganese** www.fcs.uga.edu/pubs/PDF/HACE-858-11.pdf. *Iron and manganese dissolve more readily deep underground, in the absence of oxygen. As water is pumped to the surface and exposed to oxygen, the process will reverse and the dissolved iron and manganese will precipitate out of the water forming colored sediment. Iron sediment is reddish brown or orange; manganese sediment is black or dark grey. See also* www.healthgoods.com/Education/Healthy_Home_Information/Water_Quality/iron_manganese.htm.

[29] **Non-modulating Float Valve** CLA-VAL Automatic Control Valves. 11626 Sterling Ave., Ste. F, Riverside, CA 92503. 800-942-6326, Fax: 951-687-9154. www.cla-val.com. *Highly reliable, but expensive. ½" valve costs $463, supplies 19 gpm; enough for most systems. Rebuild kit is $70.*

[30] **Watermaster** 1016 Cliff Drive #321, Santa Barbara, CA 93109. 805-966-9981, Fax 805-705-5813, usawatermaster@juno.com. *Source for ozonators, water treatment consulting, equipment selection, and installation.*

[31] **Information on sand filters** *See* www.oasisdesign.net/water/treatment/slowsandfilter.htm.

[32] **Tank Talk Newsletter** Tank Industry Consultants. *Excellent information on freezing in several past issues of this newsletter. It is geared toward large municipal tanks but much of the information transfers.* www.tankindustry.com/tanktalk.html.

[33] **National Fire Protection Association** *States sprinklers "reduce chances of dying in a fire and the average property loss by one-half to two-thirds compared to where sprinklers are not present. NFPA has no record of a fire killing more than two people in a completely sprinklered public assembly, educational, institutional or residential building..."*

[34] **Surface Water Treatment Rule** US Environmental Protection Agency www.epa.gov

[35] **National Drinking Water Clearinghouse** 800-624-8301, www.nesc.wvu.edu/ndwc. *Skilled help for small communities to run their water systems.*

[36] **Branched Drain Greywater Systems** Art Ludwig. Oasis Design. *Design, construction and use of "branched drain" greywater systems: a simple design to achieve automated, reliable subsurface irrigation without pump, filter, valves or surge tank, using all off-the-shelf components. See* www.oasisdesign.net/greywater/brancheddrain.

[37] **Republic Fastener Products Corp.** 1827 Waterview Drive, Great Falls, SC 29055. 800-386-1949, www.repfast.com. *Provides ¾" 14 gauge steel hog rings X2RP425 used for all manual hog ringing with pliers from same supplier.*

Index

Also from Oasis Design:

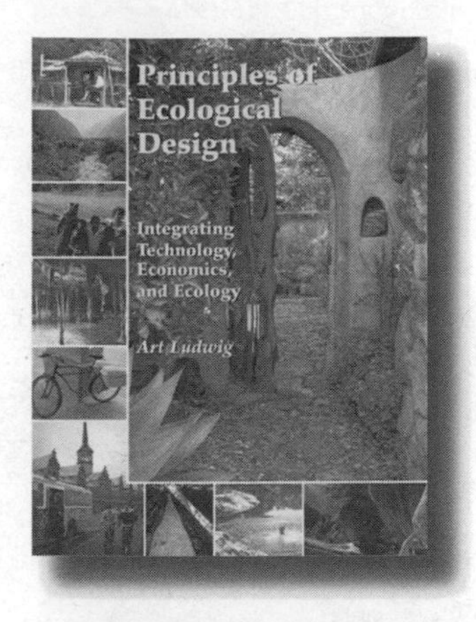

- ***Water Storage Extras***—Includes Water Tank Calculator, Research Notes on materials leaching, bacterial regrowth, disinfection byproducts, permeation, and water system component spreadsheets.
  See www.oasisdesign.net/water/storage for specs and price.
- ***Principles of Ecological Design***—Natural living is the harmonious integration of human culture, technology, and economics with nature. This booklet explains principles for redesigning our way of life to live better with less resource use. *18 pages, 30 figures and photos, all color download $4.95, or hard copy, $6.95.*
- ***The New Create an Oasis with Greywater***—Why to use/not use greywater, health guidelines, greywater sources, irrigation requirements, 20 system examples and selection chart, biocompatible cleaners, greywater plumbing principles, components, maintenance, troubleshooting, freezing, rain, preserving soil quality, storing rainwater, suppliers, further reading. *146 pages, 53 figures, 130 photos, $20.95.*
- ***Builder's Greywater Guide*** (a supplement to *Create an Oasis*)—Help for building professionals or homeowners to work within or around building codes to successfully include greywater systems in new construction or remodeling. *46 pages, 9 figures, $14.95.*
- ***Water Quality Testing Procedures and Information Packet***—A set of downloadable files which will help you learn water testing techniques which are an optimal combination of simple, inexpensive, and accurate; improve the design of your testing program; learn the limitations of testing, and how to get more out of your interpretation of water testing results. *34 page download including editable files, Beta version, $14.95.*
- ***Rainwater Harvesting and Runoff Management***—Coming soon: a new book addressing capturing runoff from a roof and using it for indoor water needs. Other applications include capturing rain runoff to flush accumulated irrigation salts from the soil, and even using runoff to generate hydroelectric power!

...and more... See www.oasisdesign.net/catalog

Oasis Design Consulting Service

A consultation is a great way to optimize your system design and ensure it fits well with your water supply, landscape, and energy systems. We can design these other systems also. Most design consultation can be done inexpensively off-site. Check our website (oasisdesign.net/design/consult) or contact us for more information.